The Dried Flower Book

first published under the title *Droogbloemen*
Diagrams by Jacques Jeuken

English language edition first published in Great Britain 1981
by The Herbert Press Limited, 46 Northchurch Road, London N1 4EJ.

Reprinted 1982, 1984, 1985, 1986

Translated by Jane Meijlink
Edited by Brenda Herbert

Printed and bound by South China Printing Co., Hong Kong

ISBN 0 906969 11 5

THE DRIED FLOWER BOOK

Growing · Picking · Drying · Arranging

Annette Mierhof

Illustrated by

Marijke den Boer-Vlamings

The Herbert Press

Contents

Hop (opposite)
Statice (below)

No so long ago, the seasons played a much more important role in the daily lives of western man than they do today. The domestic environment was influenced by whatever the different seasons could offer, and a large part of summer and autumn was spent gathering and preserving food for the winter. Hence, perhaps, the popularity of La Fontaine's fable about the ant.

Nowadays, highly advanced techniques and skilful marketing machinery make us almost unaware of the time of the year. In December all sorts of summer vegetables can be found at the greengrocer's shop, while a wide variety of fruit is imported from tropical countries. The same is true of flowers; whereas formerly only chrysanthemums or cyclamen and perhaps a few other plants could be obtained from the florist in the winter months, there is now a wide choice. Nevertheless, there is something unnatural and slightly disturbing about having cherries at Christmas and chrysanthemums in the spring.

Although it is no longer necessary to put by supplies for the winter, many of us still have an instinctive need to preserve something from summer's ephemeral beauty and abundance. The enormous pleasure to be got from harvesting and picking fruit and flowers doubltess plays its part in this.

It is pleasing to live in harmony with your environment, just as it is pleasing to live according to the seasons. A storm can be enjoyable in autumn and a January snow-fall is a wonderful sight. You can resist the winter, or you can accept it cheerfully with home-made preserves and Christmas decorations made from glossy straw fir-cones which you have gathered yourself. No carnations flown over from the Canary Islands or stock-gillyflowers from Israel can equal the satisfaction of living with what is available in your own country or your garden in the form of dried flowers, grasses and seeds.

And when your attic is hung full with bunches of dried flowers, and pots of preserved fruit and vegetables adorn the larder shelves, it is impossible not to feel a sense of pride and satisfaction. Once you become interested in the fruits of your own environment, either by growing vegetables or fruit or by collecting all kinds of plant material, your attitude to the seasons will become quite different. The way you use your garden will also change. A different corner may seem more suitable for your hydrangea, the phlox may have to make way for globe thistles or Baby's Breath. In addition, a grassy bank or woodland edge which, in the past, seemed just a confusion of greenery, will reveal itself to be full of all kinds of distinguishable plants and grasses.

Drying what you find, and growing flowers, leaves and grasses can provide enormous pleasure and become an absorbing hobby.

Where to find flowers for drying

Flowers and other plant material for drying can be found everywhere. In every kind of environment – in formal gardens, on banks and in woodland, in marshes and by the edges of pools, even by the seashore – there is something to be found which can be preserved for the winter or for an even longer period.

This book limits itself to the treatment of plants and flowers native to Northern Europe and the Eastern United States of America and similar climates. Many flowers and seeds from tropical countries can be bought in the shops but special conditions are required for their cultivation and there is not sufficient space here to go into all this in detail.

Banks, fallow land and other wild areas can provide a great deal of usable material, but your own garden will probably be the principal source of colourful flowers for drying. The average garden can produce a very considerable quantity of flowers for winter use. A selection of perennial plants suitable for drying can be included in any border in addition to a number of annuals specifically grown for this purpose. Don't be afraid that they will detract from the appearance of your garden; most of them are just as attractive as any other bedding plants. And if you refrain from picking all the flowers at once, you will obtain the maximum enjoyment from them, first in the garden itself and later indoors.

Here are a few ideas of what can be grown easily in gardens of varying sizes.

Small town garden

(see illustration on page 8)

This example contains only plants that have flowers and/or leaves suitable for drying. It is obviously not necessary to replace all your existing plants immediately, but the plan may suggest something to aim for.

New plants purchased specifically offer plenty of possibilities for drying. A garden planted with flowers for drying may look no different from any other summer garden and can be just as attractive, but once picked, these flowers have a second life. Variety can be achieved by occasionally using the bed reserved for annuals for vegetables or cut flowers instead. Harvesting the fruits from your own garden, whether flowers for drying, vegetables for eating, or simply daises, is always a rewarding experience.

Perennial border

When planning a border of perennial plants, there are many factors which must be taken into account – for example, the location, the type of soil, and the different heights, colours and flowering periods of the plants. A piece of sunny ground, preferably backed by a hedge, wall or fairly tall shrubs, is the first requirement – and a good gardening book will provide all the necessary advice and information for those who are not already experienced gardeners. The final choice of what to include, however, will depend on your personal preference.

Two examples of mixed borders containing only perennials with flowers, leaves or seed-heads for drying are illustrated on page 9.

It is a good idea to make a sketch on paper of what you are planning to put in the border. Don't be put off if you can't draw – it can be done in the form of a plan, noting the height and colour of each plant. Whether your border will fulfil your expectations or not, only time will tell. Perennials need time to show what they are capable of (although occasionally they don't even try). But the ones which flourish return faithfully every year. Although I cannot imagine my garden without annuals, it is satisfying to think that the perennials will be there year after year. Healthy perennial plants usually become fuller and more beautiful each year, and it is possible to increase the number of plants by splitting the clumps either in spring or in the autumn.

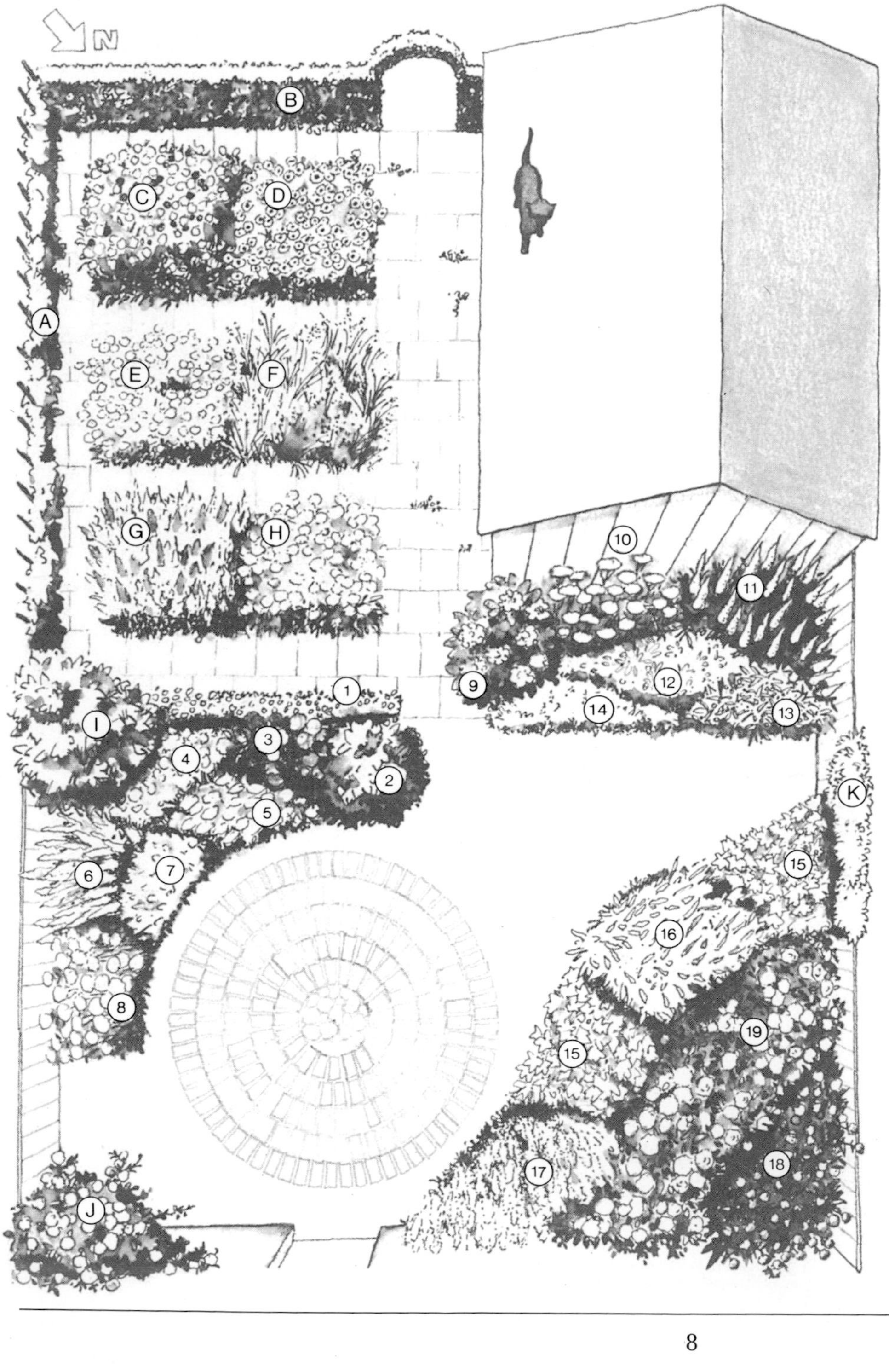

A Wineberry
B Box hedge
C Helichrysum
D Mixed acroclinium
E Rhodanthe (Helipterum)
F Grasses
G Mixed patch
F Xeranthemum
I Holly *(Ilex aquifolium f. bacciflava)*
J Climbing rose
K Clematis vitalba

1 Santolina
2 *Euonymus europaes* 'Cascade'
3 Dahlias
4 *Monarda didyma* (bergamot)
5 Achillea 'Moonshine'
6 Delphinium 'Pacific Giant'
7 *Stachys lanata*
8 Achillea 'Parker's Var.'
9 Hydrangea 'Blue Wave'
10 *Achillea millefolium*
11 *Aconitum fischeri* (low), *cammarum* (tall)
12 Artemisia 'Silver Queen)
13 *Anaphalis triplinervis*
14 *Origanum vulgare*
15 *Alchemilla mollis*
16 Lavandula
17 *Gypsophila paniculata* 'Bristol Fairy'
18 Echinops
19 Floribunda rose 'Friesia'

Suggestions for between stones or paving: *Thymus serpyllum*, Acaena

5 × 7 metres (16 × 22 ft)

Perennial border in yellow and white
Perennial border in blue and pink

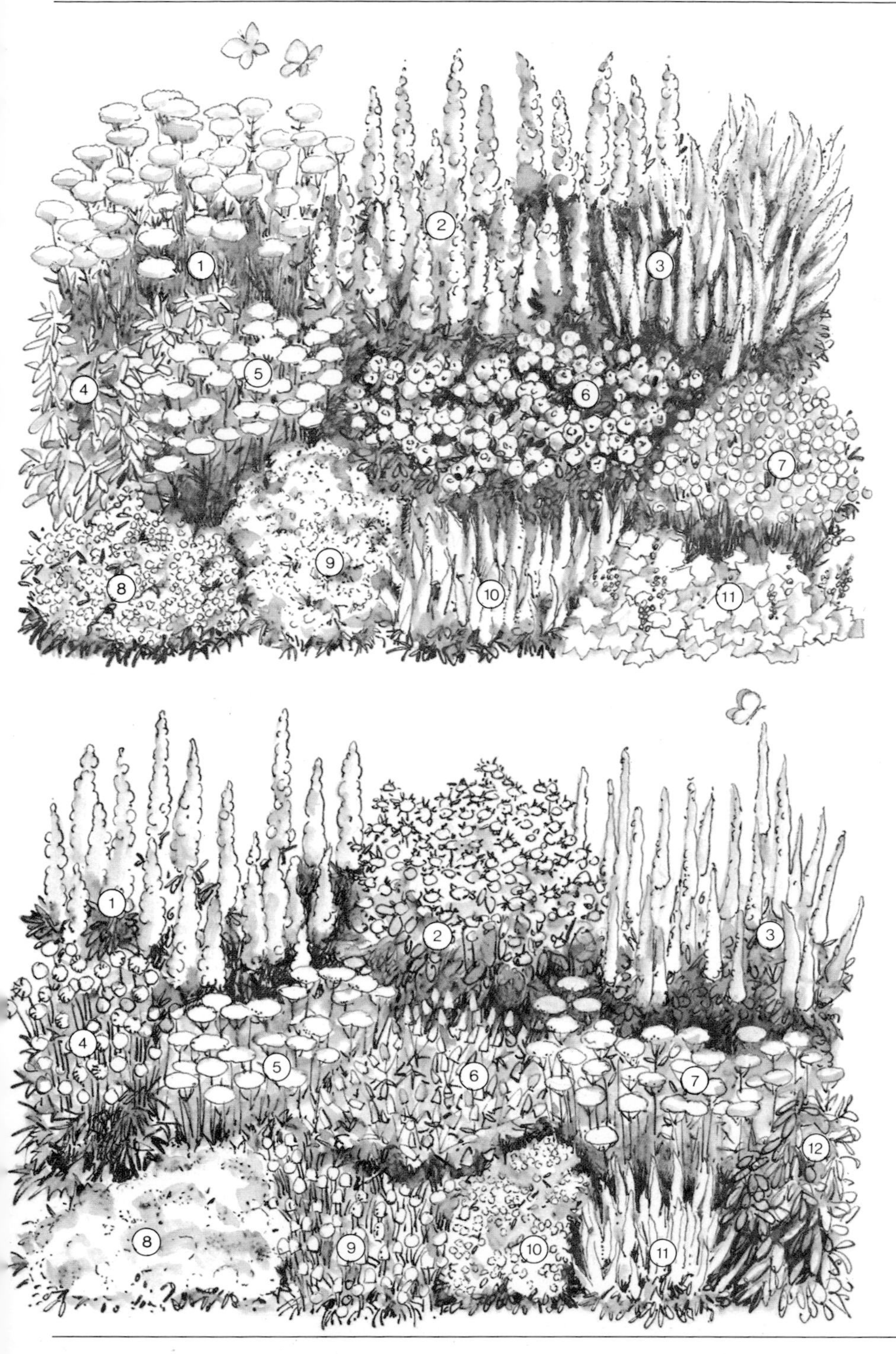

1. *Achillea filipendulina* 'Gold Plate', July
2. Delphinium 'Pacific Giant', Galahad white, June–July
3. Solidago 'Golden Wings', July–September
4. Artemisia 'Silver Queen' (grey leaf)
5. *Achillea ptarmica*, July–August (tie up)
6. Yellow floribunda rose 'Friesia', June–October
7. *Centaurea macrocephalea*, July–August
8. Anaphalis 'Schwefellicht', July–September
9. Gypsophila, July–August
10. Solidago 'Golden Gate', July–August
11. Alchemilla, June–July

1. Delphinium, June–July
2. Monarda 'Croftway Pink', July–September
3. *Aconitum carmichaelii*, August–September
4. Echinops, July–August
5. *Achillea ptarmica*, July–August (tie up)
6. Eryngium, July–August
7. *Achillea millefolium* 'Red Beauty', June–July
8. Gypsophila, July–August
9. *Armeria pseudarmeria*, May-June
10. *Anaphalis triplinervis*, July–September
11. *Lavandula augustifolia* 'Hidcote', June–July
12. Artemisia 'Silver Queen' (grey leaf)

1 Delphinium 'Pacific Giant' Galahad
2 *Achillea ptarmica* (Sneezewort)
3 Dipsacus (Teazel)
4 Lunaria (Honesty)
5 Physalis
6 Nicandra
7 Solidago (Golden Rod)
8 *Achillea millefolium* 'Red Beauty'
9 Aconitum (Monkshood)
10 Astilbe
11 Hydrangea
12 Lavender
13 Santolina
14 Monarda (Bergamot)
15 *Delphinium belladonna*
16 *Centaurea macrocephalea*
17 Achillea 'Moonshine'
18 Stachys (Lamb's Tongue)
19 Limonium (Statice)
20 Eryngium
21 Echinops (Globe Thistle)
22 Cynara
23 Gysophila (Baby's Breath)
24 Carlina (Carline Thistle)
25 Anaphalis 'Schwefellicht'
26 *Anaphalis triplinervis*
27 Polygonum affine (Knotweed)
28 *Achillea filipendulina* 'Parker's Var.'
29 Alchemilla (Lady's Mantle)
30 Buxus (Box)
31 Standard rose

Plan for a large garden for those with plenty of enthusiasm for drying flowers

a Pink rhodanthe
b Yellow helipterum
c White rhodanthe
d *Delphinium ajacis* (Larkspur)
e Helichrysum 'Feuerbal'
f Helichrysum 'Boule d'Or'
g Mixed grasses
h Lonas
i Nigella (Love-in-a-Mist)
j Silver-pink helichrysum
k White statice
l Mixed statice
m Yellow statice
n Mixed xeranthemum
o Mixed helichrysum
p White acroclinium
q Statice suworowii
r Pink statice

I Oregano
II Rue
III Chives
IV Wormwood
V Roses

10 × 20 metres (33 × 66 ft)

Basket of freshly picked flowers for drying

Any collection of material from the garden can be supplemented by wild flowers from the countryside. (Care must be taken to avoid causing damage to wild plants; but more will be said about this later.) This also means that you don't have to have your own garden in order to build up a collection. Extra supplies can be found on market stalls where a fairly wide assortment of flowers is usually available; or you can use flowers from the gardens of friends who have no interest themselves in drying flowers.

Material for drying can also be collected while you are on holiday – an attractive bunch of grasses, fantastically shaped thistles, pretty pieces of moss or unusual pine-cones. Transporting thistles, in particular, can, however, create quite a problem. I once found a number of beautiful, ready-dried, white thistle skeletons in a southern country. I picked a few and, covered in scratches, took them back to the place where we were staying. After puzzling over how to take them home, I found the problem solved for me by the concierge: the day before leaving, when I was ready to pack them, I found my beautiful thistles transformed into a bunch of gaudy prickles. As a gesture of friendship, the fellow had sprayed the dull white things bright red, green and blue. I left them behind with no regrets.

Attractively shaped leaves and ferns – which are very effective in an arrangement – are easier to transport; simply place them between newspaper at the bottom of your suitcase.

Drying flowers and other parts of plants

There are several methods of preserving flowers, leaves and seeds. But however much care you take, the result will vary from year to year. There are good years and bad years, just as with wine.

Weather conditions play an important part. A high level of humidity outside will automatically produce a high level of humidity in the air in unheated rooms indoors. The danger here is that the flowers do not dry quickly enough and so lose their colour and/or go mouldy. Whichever method of drying you use, you must use your own instinct and judgement about this.

Another important point to remember is that the length of time between picking and preparing for drying, and the amount of light received during the period of drying and storing, both affect the final result.

Try to tie and hang flowers immediately after picking.

If you want to press flowers or leaves, it is particularly important to place them between layers of blotting paper as quickly as possible. Otherwise there is a risk of the leaves curling and the flowers withering.

Direct sunlight has a detrimental effect on all kinds of dried flowers.

The colours of leaves, hydrangeas and flowers which have been dried in sand or chemicals fade especially quickly.

It is not, of course, necessary to use all the techniques described below at the same time. Even if you use only one method of drying, you can count on having sufficient material.

Magnificent bouquets can be made from bunches of flowers hung upside down. A large composition of leaves preserved in glycerine is also attractive, while very fine, small flowers from the flower press provide inspiration when you are making decorations or embellishing useful objects.

As your experience increases, you will find an incredible number of ways in which such a variety of plant material can be put together. Large bouquets with green leaves and roses, for example, or perhaps three-dimensional pictures of pressed flowers combined with all other kinds of unpressed material, can produce exquisite creations.

Drying by hanging

If possible, pick the flowers on a dry day. The exact times to pick are given in the descriptive lists of individual flowers and plants. Remove excess leaves on the lower part of the stem; they can usually be stripped off quite easily. The leaves immediately beneath the flower can also be removed, but this gives the flower a rather bare appearance; a flower with a surrounding leaf or leaves seems more natural, even when dried.

Now tie the flowers firmly in bunches and hang them up, head downwards. It is a good idea to put an elastic band around each bunch to prevent the stems falling out of the bunch as they shrink during the drying process.

The bunches should not be too large or the flowers in the centre will dry

badly, possibly turning mouldy and becoming unusable. There is also a risk of the flowers being squashed against each other and losing their shape.

Drying flat

Coarse grasses, seed-pods, mosses and even flowers can be dried flat in open boxes or on brown wrapping paper, for example.

Drying upright

Some flowers or grasses with rather delicate flowers which stand out, such as, for example, Lady's Mantle (alchemilla), the grass *Agrostis nebulosa* and umbels such as celery, parsley,dill and fennel dry better standing upright in a vase or box. In this way they can also serve as a summer arrangement. Take care not to pack the flowers too closely together or they will become entangled.

Conditions for drying

For all methods of drying in which flowers are simply exposed to the air, it

A corner of a dark attic room brightened with hanging bunches of dried flowers. In the foreground, a coarse earthenware pot filled with bunches of different varieties

is essential that the air should not be damp. Sunlight also has a bad effect. For example, drying flowers in a dark unheated attic under the roof, on the north side of the house, merely produced sad-looking bunches of mould; and a glass-covered balcony on the south side provided completely faded, straw-like bunches.

In the first instance, the air was clearly too damp; in the second the sunlight was too strong. Damp and too much sunlight are the major enemies.

Suitable places for drying are a dry attic which can be well ventilated, a staircase, or any dark corners in rooms that are well aired but not overheated. An old house with beams in the attic is, of course, perfect for drying flowers, but an ordinary washing line hung in any none-steamy room is also suitable (i.e. not the kitchen or bathroom). If the bunches are hung carefully, with some attention to colour, they form an attractive decoration while drying.

It is important to find a location where the temperature can be regulated; this is largely a question of common sense. A bad example in this respect was the florist whom I telephoned once in October for some dried eryngium. He would be pleased to help me, he said, but his flowers were unfortunately not yet dry; further questioning revealed that they had been hanging in a closed attic with no form of heating during a very wet summer. The stems were still soft and many of the flowers were mouldy.

If, after hanging for a few days, the

A selection of leaves which will dry well between layers of newspaper:
1 Pasque Flower 3 Japanese Maple
2 White Poplar 4 Wineberry
5 Southernwood

stems do not feel appreciably drier, some form of heating is desirable to dry the air. It is more a question of humidity than of temperature. To be really professional about this you could use an hygrometer, keeping the humidity at 40–50%.

There are certain plants such as delphinium, mullein (verbascum) and purple loosestrife which should preferably be dried quickly. An airing cupboard is ideal for this purpose, but it is possible to make do with the boiler cupboard or the corner reserved for the central heating unit. Once the flowers are dry, they can tolerate rather more humidity. Over-dry flowers are difficult to arrange; they become brittle, breaking off at the least touch. Flowers which are too dry are impossible to mount on artificial stems.

Pressing

All leaves can be dried by pressing. Try to collect a variety of shapes and colours.

There are two distinct groups of material for pressing. The first consists of large leaves and ferns to be used in arrangements. In the second group are small leaves and delicate flowers which can be used to decorate Christmas cards or objects such as boxes and trays, for example. Instructions for pressing flowers are given in a separate chapter (page 85).

The first group does not need a great deal of pressure, or the material will become too stiff and brittle and therefore difficult to arrange. Since leaves should be able to breathe, heavy pressure is quite wrong. A good method

More leaves which can be dried between newspaper:
1 Columbine 3 Cineraria maritima
2 Sea Wormwood 4 Polypody

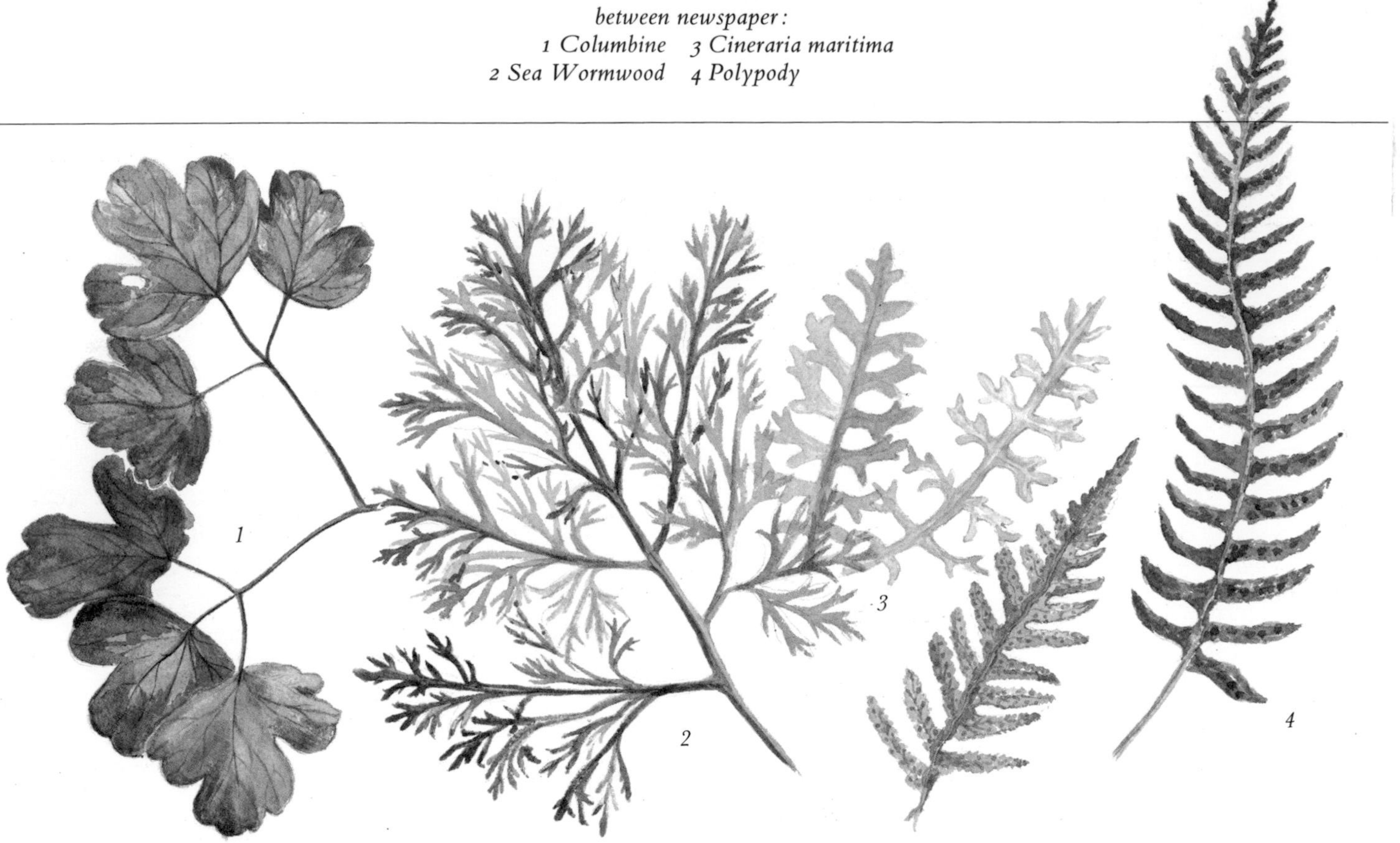

is to place the leaves, ferns, etc. between sheets of blotting paper or old newspapers and then slide them under a carpet. Alternatively they can be placed under a mattress or under loose chair seats. It is important to make sure that the leaves are completely dry; drops of water will leave ugly brown marks. To avoid curling, place them between the sheets of paper as soon as possible after picking. Drying time depends on the thickness of the leaves and the paper used, but a week is usually sufficient.

A large quantity of material can be collected in a short time, using this method, particularly as the sheets of paper can be stacked on top of each other. Leaves dried in this way can be kept for years.

Some leaves recommended for drying

Ferns
Grey-tinted leaves such as senecio or cineraria, Wineberry (grey on the back), grey-leafed poplars
Japanese maples (*Acer Japonica*): attractively cut, red-coloured leaf
Strawberry leaves, just fading
Box
Acanthus
Alchemilla mollis (Lady's Mantle)
Oregon grape
Elaeagnus angustifolia (Oleaster)
Conifers remain in good condition for some time but eventually become brittle.

The glycerine method

This method is, in fact, a type of preserving and has little in common with drying. It is generally used for coarser leaves such as magnolia, rhododendron and beech. A list of suitable material can be found at the end of this section. To reach a state of preservation the leaves and stems which have been picked must absorb a mixture of glycerine and water. To ensure thorough absorption, woody stems should first be beaten flat or split.

The leaves to be used should be fresh and fully grown, which means they should be picked at the end of summer or in autumn. If you want autumn shades pick them just as they are beginning to turn colour. Do not use leaves which have completely faded, as they will not absorb glycerine. This method of preserving tends to cause the leaves to fade and become browner; but they remain supple and are therefore easy to arrange, and often very attractive shades are produced.

Storage presents no problem, as the treated leaves will remain in good condition if placed in boxes under the bed or in plastic bags. Excessive damp can cause mould.

A few mothballs, or a bag of dried wormwood and southernwood, will keep intruding insects at bay.

Glycerine method 1
Add two parts boiling water to one part glycerine (obtainable from any chemist) and mix well. Place the stems (split if necessary) in the solution, preferably in a narrow pot, so that they are covered about 7–10cms (3–4ins) deep. The solution should still be warm. Before placing the stems in the glycerine, it is a good idea to dip them for a minute or two in plain hot water to set the sap circulation in motion.

When drops of glycerine form on the leaves, enough has been absorbed and the stems can be removed. Leaves which have absorbed too much glycerine will droop; excessive oil should therefore be wiped off.

The uppermost leaves of tall branches or sprays are sometimes not reached by the solution and will then wither. Since the whole effect of the branch is then lost, it is better to use the second method, described below.

Glycerine method 2
A long flat container is needed so that the whole branch can be submerged.

The mixture consists of equal quantities of glycerine and water. The container must hold sufficient liquid to cover the leaves without their being too closely packed. Small stones can be laid on the leaves to hold them under.

The length of time necessary to preserve the leaves varies from plant to plant. As a general rule, a branch can be removed from the solution when two-thirds of the leaf has become darker and more transparent than the remainder. Any leaves which are still green and not transparent must remain in the liquid.

When they are ready, place the leaves to drain for a few days on a pile of unopened newspapers. The top part of the leaves will continue to change colour as they dry. They will become supple, and will remain so. After a few days the leaves should be carefully washed with a little soap and water and then rinsed with clean water and pegged on a line to dry. If required, these leaves can be sprayed with green paint. Leaves steeped too long in the glycerine bath become too limp to use in arrangements.

Only fully mature leaves and those that are not too delicate should be used for this method. Ferns will immediately curl and become unusable.

The dry branches should be stored as described above.

I have had little success in preserving flowers by this method. Only *Molucella laevis* (Bells of Ireland) and common heather – which should not remain in the solution (method 1) longer than three days – tend to hold their shape better when preserved in this way. If heather is left longer in the glycerine, it will certainly fade.

The following is a selection of leaves suitable for preservation in glycerine: Beech leaves, most grey leaves, holly, oak leaves, dark-leafed prunus, Japanese maple, ivy (only in glycerine bath), *Hydrangea paniculata*, heather, paeony leaves.

Drying with sand and other water-absorbent material

Of all the methods so far discussed, this is the most fascinating. It has great possibilities. Geneal Condon, an American expert, has written a book about the science of preserving flowers in sand and her experiences in this field. She uses sand alone to draw the moisture very slowly out of the plant, which is completely embedded in the sand. This method was also formerly used in India to preserve roses.

Unfortunately, I have had no success with this method – maybe because, in the first place, Mrs Condon used a special type of sand from the desert near Salt Lake City and, secondly, the climate in parts of America is quite different from that of north Europe.

There is, however, another method of working with sand which has proved successful and which is described below.

The colours of flowers dried in sand can best be preserved by artificial means – for example, by brushing with powdered chalk, with a marten-hair brush, or by painting the flowers when dry. But this is not essential. Excellent results are generally obtained without resorting to such tricks. This method is very rewarding, especially after a little practice to reduce the chance of failure.

There is still one problem: occasionally flowers which have emerged flawless from the sand suddenly droop a few weeks or months later. This is probably due to the humidity in the air, which affects the firmness and the colour of the dried flowers. There are fortunately several varieties, listed on page 20, which are less likely to be affected in this way.

Flowers which droop do not necessarily have to be thrown away. A slightly crumpled example can still find a place in an arrangement where variety in shape and colour is essential. There is no real solution to the problem. A little hair-spray can give some protection. On the other hand, you may be lucky enough to have no damp problem in your home at all.

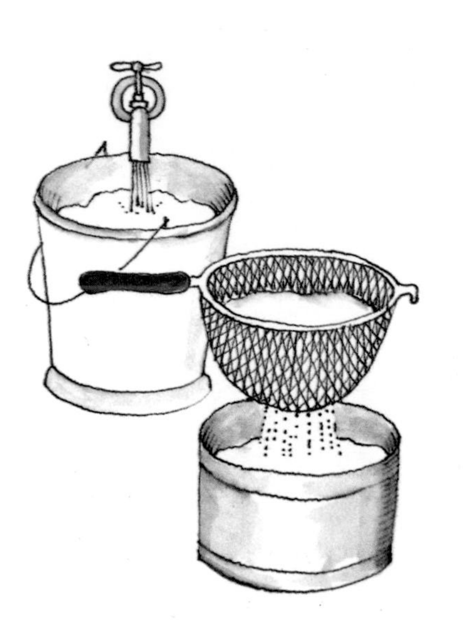

Materials for drying with sand

Sand

In principle, any fine-structured sand can be used. The best and easiest to use is white silver sand which can be obtained from garden centres and some hardware stores. Other kinds of sand must first be washed thoroughly a few times. To do this, place the sand in a bucket, pour on water and stir well so that any impurities can float to the top; leave the sand to settle and then pour off the dirty water. Repeat the process until the water looks clean when the sand has settled. Spread out the sand and leave to dry.

The same sand can be used repeatedly. If, after some time, it becomes contaminated with particles of plants and flowers, it can be cleaned simply by sieving.

Silica gel

These are water-absorbent crystals which can be obtained from a chemist or drug store. When absolutely dry, the crystals are blue; when they absorb moisture from their surroundings, they turn pink. Damp crystals can be dried and turned blue again in the oven or even in a saucepan. Only completely dry silica gel should be used for drying flowers.

Containers

Anything which can be sealed to exclude air is suitable for this method of preserving flowers: for example, tins or plastic boxes. It is useful to have several containers of various shapes and sizes. Empty plastic containers from supermarkets, or cake tins, are ideal; but cardboard should not be used in a damp climate as it absorbs too much moisture.

If containers are to be placed in an oven or heated area, they should not be closed: as the air outside the tin will be drier than that inside, the drying process would merely be retarded.

Plant material for drying in sand

Almost any kind of plant material can be dried in sand, although there are certain important points to remember. The leaves and flowers must be placed in the sand as fresh as possible. If they are left for any length of time, they become limp and change shape.

The flowers must also be absolutely

Roses placed in a box of sand

dry. Any drop of water which passes unnoticed can cause ugly brown spotting; or, if the centre of the flower is damp, it may start to rot, causing the petals to fall off.

You can check that the flower is completely dry by carefully sprinkling it with sand. If the sand does not adhere anywhere, everything is in order.

During a particularly rainy period, the flowers can be dried off first indoors in a vase. In this case they should be picked while not yet fully open, or they will be likely to disintegrate during the drying process.

Method

Although sand may feel dry, it can still contain too much moisture for the drying process; the sand may run smoothly through your fingers, but the final result will be withered, brown flowers instead of well-dried ones.

You can use the silica gel as an indicator of any possible moisture present in the sand. If the blue silica gel crystals change colour when mixed with the sand, the moisture content is clearly too high. Since large crystals cause spotting on the petals if they come into contact with the flowers, it is better to sieve out the largest first, leaving the grit which can be mixed with the sand.

The mixture should now be placed in the oven at 300°F (150°C). After some time the blue silica gel will be clearly distinguishable from the white sand which is now dry. (The remaining large crystals can be dried in the oven at the same time in a separate container.) The silica gel grit in the sand will speed up the drying process and cause no harm to the flowers.

Put a layer of sand mixed with silica gel grit in a container and place the flowers on it, being careful to arrange them so that they keep their shape. Flowers dried by this method usually have to be mounted on artificial stems after drying, so 1–2cms ($\frac{1}{2}$in.) of natural stem should be left for this purpose. It is best to place the flowers on the bed of sand face upwards; but if the container is too shallow, this may not be possible and they will have to lie slightly on one side.

Although several flowers can be placed in one container, they should not touch each other. Never place two different varities in one box: drying times vary, so one flower may emerge beautifully dry while the other is limp and membranous.

A piece of fine chicken wire placed just above the base of the container and covered with sand will hold the stems, inserted vertically, so that the flowers remain in position. If the flowers are to be dried on their sides, take care that the sand reaches right inside the flower or the centre may rot. To fill the flower with sand, make a ring with your thumb and index finger on which to rest the flower while you gently trickle sand into it with the other hand.

When the flowers are in position, they and the area round them should be lightly sprinkled with sand until they are completely covered. During this process, neither the shape nor the position of the blooms should change. If necessary, any slight movement can be corrected with the handle of a fine paint brush.

If the flower heads are damp, or of the double variety, a few of the larger silica gel crystals can be sprinkled around them but must not touch them.

It is also a good idea to sprinkle a layer of silica gel on top of the sand before closing the box.

Place the closed containers in a really dry, fairly warm place. Flowers placed on a radiator which is turned on occasionally, as in early autumn, for example, will usually dry in three to four days.

Drying can also be carried out in the oven at *c.*110°F (40–50°C); twenty-four hours is usually sufficient, though this will depend on the variety. The short drying time is an advantage because it allows you to dry a large quantity of flowers in one season, using little sand. But it does use up a lot of gas or electricity which is more or less wasted in the process.

After about five days (if not using the oven), feel carefully in the container with one finger. If the first petal you touch feels dry and rather crisp, it is time to tip the sand gently out of the box. This must be done with care as the dried flowers are brittle and will break easily. Pour the sand through a sieve to remove the silica gel crystals. Any sand remaining on the petals can be removed with a small paint brush.

It is best to mount the flowerheads on stems immediately. At this stage the receptacle (where the stem joins the flower) is often still soft enough to be perforated easily, and with many varieties (roses, dahlias and zinnias, for example) an artificial stem can be made by passing florist's wire through the receptacle and up through the centre of the flower. A small hook should be made at the protruding end of the wire and then carefully drawn back into the flower until firmly lodged in position. This hook must, of course, be small enough to be concealed within the flower centre.

Do no wait too long before inserting

the wire; if the receptacles have had time to harden, this can present difficulties.

One of the major problems with this method of drying is, once again, storage. Humidity must be avoided; and flowers laid on their sides in boxes lose their shape. The best solution is to insert the wire stems in a block of

Oasis and store them in this way in a very dry, dust-free place. The flowers can also be tied in bunches and hung up; this is ideal for roses, for example.

Although all flowers can be dried in sand with silica gel, some varieties are more successful than others due to their good keeping qualities. Double flowers generally keep better than single ones. In a humid atmosphere, single flowers soon droop, whereas the petals of double flowers support each other.

If you want to dry composite flowers such as daisies or marigolds they must be picked when the outer ray of petals have just opened. If you wait too long, the outer petals will fall off when you remove the specimens from the sand. In all other respects this type of flower dries beautifully.

Some types of flower suitable for drying in sand

Roses – 4 days
Delphinium 'Pacific Giant' – separate heads – 3 days
Larkspur annual – separate heads – 3 days
Astrantia – 3 days
Hollyhock (double) – 5 days
Pompom dahlia – 5 days
Zinnia – 3 days
Ground elder – 3 days
African marigold – 5 days
Hortensia – unfaded blooms – 3 days

Try also calendulas, helichrysum, chrysanthemum, paeony and perennial scabious.

Sand-dried rudbeckia (Black-eyed Susan)

There are other chemicals besides silica gel which can be used in the same way: for example, borax and alum powder. The disadvantage of these products is that they tend to turn lumpy. It is also rather difficult to tell if they are absolutely dry as they do not change colour when damp.

A final tip

If, for some reason, short-stemmed, recently picked flowers have to be kept for some time before drying, take a wide, shallow dish and fill it with water. Cover this with a piece of aluminium foil in which holes should be pricked to hold the flowers so that the heads remain dry above the foil while the stems can absorb water. Alternatively the flowers can be inserted in a block of very wet Oasis.

A basket arranged with bunches of different kinds of dried annuals and filled in with wild flowers and grasses

Annuals

The most popular assortment of dried flowers is generally composed largely of annual flowers. They are not usually difficult to grow, certainly no more difficult than the average African marigold or petunia. They flourish, moreover, in virtually any type of soil. Although the method of sowing is generally given on the seed packet, it is important to note the following points.

You will achieve the best results if you sow the seeds in the spring in small pots or seed trays and later, when there is no further danger of frost, transplant them into the garden. This sounds simple but the chance of failure is unfortunately considerable. You would be well advised to consult one of the many books available on growing annuals to learn what should be done; I shall restrict myself here to a few general words of advice.

When the seeds I have ordered arrive (about half way through March), my window sill soon becomes covered with pots and containers in all shapes and colours, covered with sheets of plastic or pieces of glass, in which everything is sown.

After a few days, some of the containers will already be covered with an attractive flush of green while others may take several weeks to germinate.

You should never dig the seeds up if nothing is visible. They will usually have already grown their first shoot which will immediately break off, reducing the chances of survival.

Seeds which have already germinated will now slowly grow into thin blades with two small green leaves at the tip; these will try to reach the generally rather restricted amount of light. There is a considerable chance that by the following week they will all be growing at a slant in the direction of the window. If this happens, something is wrong – the window-sill may not be the most suitable place, because the light comes too strongly from one direction. It may also be too warm; 50–60°F (10–15°C) is sufficient.

If you haven't a suitable place indoors, it would be better to sow the seeds outside in a small greenhouse or cold frame. But you must wait until the beginning of April, since there is then more light and less chance of frost. If frost is nevertheless forecast, don't forget to cover the plants over or they will be black and dead the next morning; you can use newspapers or large sheets of plastic. If you don't have a greenhouse, the containers can be covered with a sheet of glass and placed in a light, sheltered spot.

Avoid too much sun, which could shrivel up the seedlings. The containers should be well ventilated, since condensation can cause rotting.

Once you have succeeded in bringing the seedlings through the first delicate period, the next stage is to transplant them. This can be done when the plants have 4–6 leaves and can be handled easily.

One recurrent problem is how to find sufficient space in which to plant out all those thousands of seedlings. The only answer is to give them a good home by passing them on to friends and neighbours.

If the weather is co-operative, you will soon be able to plant them in the garden. With a bit of luck, you can choose a mellow, rainy day in May. The rain makes the earth smell of all the compost you applied in early spring.

The flowers are usually planted out in beds kept for annuals, but you can also plant them in a mixed border.

You can start planting out of doors from around the third week in May but you should keep an eye on the weather forecast since a hard frost can still cause untold damage. The damage can be limited by walking around with pieces of smouldering turf to set up a smoke screen, and thoroughly spraying with water, but this has to be done in the middle of the night when the frost strikes. Not everyone will want to carry their enthusiasm for plants to such lengths.

This anxious stage is followed by a period of calm: a little weeding between the rows from time to time, perhaps spraying with water now and again when the ground is dry, and the rest will take care of itself.

The flowering periods for different types of annuals vary widely and will be mentioned in the list of flowers which follows. You can begin to pick from July onwards and continue until the first frost. One morning in October, the plants will be black and limp – they are then ready for the compost heap (and not, as is so often seen, for the rubbish heap).

In the meantime you will have hung your attic or staircase full with the colours of a beautiful summer.

Acroclinium roseum or Helipterum roseum

These flowers somewhat resemble daisies in form. Some are white with a yellow centre, others pink with a yellow centre (I find these very difficult to combine with other flowers), or there are some which are almost red with a black centre.

The first flowers can often be picked at the beginning of July while new ones are still growing. The flowers can be gathered in full bloom, but it is nice to pick a few still in bud.

If the weather is damp, the outer petals will close over the hearts. If, due to continual rain, for example, you have to pick them while they are closed it presents no problem: they will open up again later when indoors and dry. If the flowers remain too long on the plants, they soon spoil.

Strip off the lower leaves and hang up the flowers to dry.

Amaranthus (Love-lies-bleeding)

Rather attractive flowers on long spikes, erect or drooping, in shades of red or pale green.

The plants are simple to grow and easily propagate themselves again.

The time to pick them is somewhat arbitrary as the flowers do not change very much during the flowering period.

The spikes can be hung up to dry.

Ammobium (Winged Everlasting)

This plant is in the form of a rosette from which grow tall, fleshy stems bearing small, white flowers with yellow centres which often turn black when dried.

The seeds germinate very easily; they can be sown in open ground at the end of April but should be well thinned out.

Pick the flowers when the yellow centre is just visible.

If you want completely white flowers, pick them before they open. The time to pick is from mid-August onwards. The flowers are most effective in arrangements when tied in small bunches.

They should be hung up to dry.

Delphinium ajacis (annual larkspur)

Two varieties can usually be found in catalogues: the Giant Imperials and the hyacinth-flowered strains.

Colours: white, pink, blue and purple.

Both kinds can be used for drying.

Cultivating them is not always so simple. The packets state that they can be sown directly in the open ground. But since the seed often tends to come up badly, it is better to sow half in a tray and later use these to fill up the bare patches in the bed.

They can be picked in August when the flowers are fully open. Do not wait too long or the flowers will tend to fall off during the drying process.

The hyacinth-flowered varieties, in contrast with the Imperials, have no side branches so the plants can be lifted immediately after picking.

Hang up to dry.

Gomphrena globosa (Globe Amaranth)

Usually purple but sometimes also pink and white globes. They are often sold as pot plants.

These rather small 'buttons' are best arranged in bunches. They should be picked from mid-August onwards. The timing is not very important as their shape and colour change very little.

Hang up to dry.

Helichrysum

This is the true Straw Flower, often rather stiffly used in run-of-the-mill dried arrangements, usually mounted on a wire, together with an ear of corn.

Nevertheless this flower is not to be despised. It is available in many colours: a good seed merchant can usually offer eight colours.

It is easy to grow and blooms right into the autumn. The first flowers appear around the end of July; these should be picked carefully as the plant will bloom again, usually with slightly smaller flowers. Pick the flowers when the centre is not yet visible as they will open further during the drying process. It is a good idea to pick some still in bud. This helps to emphasize the natural effect in an arrangement.

After picking, remove the lower leaves. Bare stems can be tied more easily. Tie the string securely, as the stems shrink during drying and the bunches will otherwise fall apart. The best solution, of course, is to use elastic bands instead of string.

Don't remove the upper leaves or the flowers will tend to look very bare when arranged.

Helipterum humboldtiana (= H. sanfordii)

Attractive pure yellow umbels which retain their colour for years. Growing these flowers can, however, prove a disappointing experience. In any case, they need very fertile soil; but the flowers may still be very meagre and the reason for this is not always evident.

If all goes well, this variety is one of the earliest to be picked, often at the end of June. New flowers will grow after the first picking but they will be rather smaller. Do not pick until the small flowers which form the umbels are fully open.

The flowering period is short, and the plants can usually be lifted in August.

Hang up to dry. (See also *Acroclineum roseum*, page 24 and Rhodanthe, below.)

Helipterum manglesii

This flower is also known as Rhodanthe, and is sometimes confused with *Helipterum sanfordii* which it in no way resembles. The rhodanthe has small, pink or white flowers with silvery buds; the centres are yellow. It likes a humus-rich soil.

Along with the *Helipterum sanfordii*, it is one of the earliest flowers to be picked – at the end of June. Pick when a few flowers are open on a stem, since the glossy silver buds are by no means the least attractive aspect of the plant.

The remaining stems will form no new flowers so you can lift this plant early.

Hang up to dry.

Lonas inodora (African Daisy)

These are dark yellow umbels consisting of tiny bud-like flowers. Success is certain with this variety. The strong plants provide flowers right up to the first frost. Here, too, the first flowers are larger than those formed later.

Once the flowers are fully open, the time to pick is not important; they do not change very much during the blooming period. Can be picked from August onwards.

Hang up to dry.

Molucella laevis (Bells of Ireland)

A less well-known plant with a rather insignificant, small white flower surrounded by an attractive cup-shaped, pale green calyx. The stems are covered with these little bells.

Flowering time is late summer. Pick when all the bracts are open and rather stiff to the touch.

Hanging up to dry works well, but the glycerine method can also be used. The flowers should not be dried too quickly as they tend to fall apart rather easily.

The delicate, pale-green colour changes quite quickly to beige under the influence of light.

Bells of Ireland can also be used very

decoratively in summer arrangements of fresh flowers.

Nicandra (Apple of Peru, Shoo Fly Plant)

When sowing nicandra in your garden for the first time, you will be introducing a new, but friendly, weed. It is a large plant which blooms with pale blue, bell-shaped flowers.

For drying purposes, it is the seed-pods which are of interest. These seed-pods are protected by lanterns which vary in colour from nearly black to pale gold-green.

Apart from their use in all kinds of arrangements, they are also excellent in Christmas decorations, especially if lightly sprayed with gold. They can be picked from September onwards.

A few plants can be left standing; they will provide a sufficient harvest the following year.

Nigella (Love-in-a-Mist)

The most romantic of all dried flowers. Apart from the blue variety, there is also a hybrid called Persian Jewels. The seeds can be sown directly in open ground. The beautiful blue flowers are surrounded by delicate, lacy, sprigged green bracts.

The flowers can be very successfully dried in sand; but the main interest lies in the seed capsules which are similar in shape to rose buds and still surrounded by that lacy 'mist'.

It is best not to pick until all the seed capsules have been formed. The entire plant can then be pulled out of the ground. The roots should be snipped off, leaving ready-made bunches.

Hang up to dry.

Physalis Peruviana var. edulis (Cape Gooseberry)

This plant really belongs in the kitchen garden. A few years ago I sowed this variety of lantern for the first time. I hoped to be able to pick the fruit in the autumn but unfortunately the north European summer is often too poor for the fruit to ripen. But even if the berries are inedible, the plant produces plenty of fine, pale green lanterns which are excellent for drying: they retain their attractive pale green colour, which enlivens dried boquets and gives an impression of fresh flowers.

An interesting effect is produced by cutting the 'lanterns' open four or five times to form a star-shaped green flower with a round cherry in the centre. And who knows, some may even ripen! If not, buy a tin of golden-berries to sample their taste. (See also the perennial *Physalis franchetii*, page 37.)

Statice (limonium) sinuata

A very well-known flower for drying which can be obtained nowadays the whole year round from florists and market stalls.

Statice forms umbels of small, deep pink, white, yellow, blue and purple flowers. The most commonly available is the purple-blue variety which in my opinion is the least attractive. If you buy a packet of seed of mixed colours, you may be in for a surprise – for example, a very pale lilac or a beautiful yellow-pink combination. 'Pacific Giants' is a particularly good mixture for cutting and drying.

These flowers make an excellent basis for a bouquet.

They need plenty of sunshine while growing. A rainy summer reduces the harvest to a pitiful minimum, and the flowers may rot before they even open. Pick as many as possible on dry days when most of the flowers on the umbels are open. The plant continues to form new flowers until the first frost.

Dry hanging up. To avoid mould and spoiling the shape of the umbels, the bunches should be kept small.

Statice suworowii

This variety, with its long purple-pink tails, is also well known. The spikes are covered with very small flowers.

Cultivation is not always easy; this statice also needs plenty of sunlight. The young seedlings must be transplanted on time if you want strong plants.

Always pick with care since you may otherwise pull the entire plant out by mistake.

Some annual flowers to be grown from seed:
1 White rhodanthe (Helipterum) 4 Statice suworowii
2 Nigella (Love-in-a-mist) 5 Pink acroclinium (Helipterum)
3 Nicandra 6 Delphinium ajacis
7 Pink rhodante

The first spikes are the largest; later ones will always be somewhat smaller, but don't expect to see too many. A lot of plants will be needed to obtain a few, reasonably-sized bunches.

They flower rather early, around mid-July.

They are very decorative in bouquets and you can use them to outline the shape of the arrangement. They should be hung up to dry.

Xeranthemum annuum

White and purple are the colours that occur most commonly. Shades of pink and lilac can be expected from packets of mixed seed. The flowers are rather star-shaped, borne on stiff stems which make them easy to use in bouquets. The white variety is very attractive when used in Christmas decorations, Christmas wreaths, Christmas balls and pyramid trees (see page 75).

A single plant can provide an abundance of flowers. They can be picked from the appearance of the first flower in August until the frost puts an end to the prolific supply. Pick the flowers when fully open, but also pick a few in bud. In large arrangements, they are best used in bunches.

Hang up to dry.

This 'triangular' bouquet contains many sand-dried zinnias. Their colours are repeated in the prunus leaves and the dark red hortensias.

Perennials

Drying perennials can often produce surprises. It is quite likely that the following list includes plants which have been in your garden for years but which you never considered as flowers for drying.

If your garden is increasingly dominated by perennials grown specifically for your dried flower hobby and if they are, moreover, stripped bare at the very moment when they are blooming at their best, this could lead to something less than enthusiasm from the other members of your household.

A little forethought is necessary.

If sufficient space is available, additional flowers can be planted so that it will not be necessary to pick everything at once.

The following summary of selected perennial plants includes little or no information about growing methods, location and type of soil. This information can be found in good gardening books or growers' catalogues. Unless otherwise indicated, the flowers should be picked in full bloom, in dry, sunny weather and hung up to dry (see also page 13).

Acaena repens

An attractive rock plant which is also suitable for growing between the paving of a terrace or path. It has a decorative leaf and brown-red fruit. Pick when the burrs are a beautiful shade of red (August).

Flowering period: June–July.

Achillea (Yarrow)

Most people know that the flowers of this plant can be dried. The best known variety is *A. filipendula* 'Parker's Variety' with mustard-yellow umbels.

Flowering period: July–August.

Hybrid 'Moonshine': a smaller variety of Achillea with a grey-green leaf and bright yellow umbels.

Flowering period: June–July.

A. millefolium 'Cerise Queen': all shades of pink to crimson. This variety of achillea has parasitic roots and is therefore more suited to some untidy corner of the garden.

Flowering period: July–August.

The varieties of achillea mentioned above should be picked when the umbels have grown to an attractive round shape and each single flower is in bloom.

A. ptarmica 'Perry's White' is a variety of achillea with a quite different flower, with double white buds. An attractive flower for drying, which can best be used tied in small tufts.

Aconitum (Monkshood)

A plant with a descriptive name. Attractive single raceme in various shades from pale violet to purple-blue.

The fact that the hybrids vary in height and flowering period provides interesting possibilities for combining them in borders.

Flowering period: July–September (depending on the location, which should not be too shaded).

Note: all species of aconitum are poisonous.

Alchemilla mollis (Lady's Mantle)

A plant which suits every kind of garden, and which looks particularly attractive with a few drops of rain on its velvety leaves. It has delicate, yellow-green umbels which should be picked when the tiny star-shaped flowers are wide open.

Flowering period: June (sometimes blooms again in the autumn).

Allium (decorative onion)

A very extensive plant genus, some species of which are good for drying. A useful tip: all compact flowerheads in bloom are suitable.

A. schoenoprasum: better known as chives. This plant can be used in two ways: as a herb to season food (pick leaves before flowering) and as an attractive, pale lilac, dried flower (pick before fully open).

A. giganteum: as its name indicates, this has enormous stems with purple flowers suitable for large bouquets.

Flowering period: July–August.

Pick when the flowers are open all the way round.

Eryngium planum

Perhaps out of place with perennials is the leek. If you have a vegetable garden, grow a few leek plants there. They will flower the following year in July.

Anaphalis margaritacea (Pearl Everlasting)

Ivory-white flowerheads, which should be picked before they are fully in bloom. The fluffy centre of the flower fades to yellow-brown. If this happens and you want to revive the colour, you can pluck out the centre (using a pair of tweezers). This job is not recommended for those suffering from hay fever and allergy, but the result is a surprising grey-green receptacle.

Flowering period: July–August.

Hybrid 'Schwefellicht': bright yellow flowers and, like the varieties described above, beautiful silver-grey leaves which should certainly not be removed. Pick when the deep yellow centre is visible.

Flowering period: July–August.

Armeria maritima 'Vindictive' or Armeria pseudarmeria (Sea Pink or Thrift)

An old-fashioned-sounding name which recalls gardens with gravel paths edged with these plants. Commonly found wild along the sea coast. Easy to grow from seed.

Pick before all the crimson-pink flowers are open.

Flowering period: June–July.

Artemisia ludoviciana 'Silver Queen'

Stems with attractive, pointed, silver-grey leaves.

Silver Queen blooms very modestly at the end of the summer season; do not pick until the branches have formed side shoots.

Other species of artemisia, for example wormwood, also dry well. In addition these plants repel moles, mice and other vermin.

Astilbe arendsii hybrid

A plant which is better known (incorrectly) as Spirea.

If you want to give this plant a chance in your garden, sow it in a damp place (excellent next to a pond). Varying shades of pink and red.

Flowering period: July–August.
A. Chinensis: deep pink variety of astilbe which can grow on slightly drier soil.

Flowering period: July–August.

Astrantia alba, Astrantia rubra

The white-green (*alba*) or deep pink (*rubra*) flowers resemble miniature Biedermeier bouquets.

Flowering period: June–August. They should be picked when the stamens are visible. If you hang them to dry, they will close up a little, but the colour does not change.

Astrantia will be even more attractive if dried in sand.

Calluna vulgaris (Heather)

Innumerable varieties with flowers and foliage in all possible colour nuances.

The flowering period varies considerably according to the species.

Pick when the flowers on the racemes are for the most part open.

If sprayed with hair lacquer they will disintegrate less quickly. They can also be placed in glycerine for a few days (see page 17).

Centaurea macrocephalea

This centaurea resembles a thistle more than a cornflower. The flower looks like a yellow powder puff. Pick in full bloom: it will remain a beautiful yellow colour. If you cannot bring yourself to pick such a unique phenomenon in all its glory, wait until the flowers have finished blooming; the receptacle resembles a glossy, golden daisy.

Flowering period: June–July.

Cynara scolymus (Artichoke)

The choice is yours: eat it or dry it!

If you decide to dry it, the buds should be left to grow until just open. The purple flower is suitable for larger arrangements.

Flowering period: July–August.

Similar flowers, but much smaller, can be obtained from cardoons.

Delphinium 'Pacific Giant'

These flowers lend themselves well to drying. The side branches formed after the main stem has been cut off are somewhat thinner and can be used in small bouquets.

The colour variations range from pale blue to purple, sometimes with a dark centre.

They should be picked when the racemes are nearly all open.

Flowering period: end June–July.

When dried, the colour (especially of the dark blue varieties) is retained for years.

Dipsacus fullonum (Teasel)

This cultivated thistle (biennial) is a good substitute for the wild variety, which is not always easy to find.

If you pick the heads before they bloom, the colour will remain lighter.

Flowering period: July.

Remove as many leaves and prickles as possible immediately; as the plant dries the prickles grow sharper.

Echinops banaticus (Globe Thistle)

This should be picked before it actually flowers, when completely blue but before the tubes can be seen. (Thistles picked too late will soon disintegrate when dried.)

During the summer months florists sell fresh bunches of globe thistles (sometimes combined with yellow achillea) which, for some reason unknown to me, are always much bluer than mine – so I simply buy a bunch.

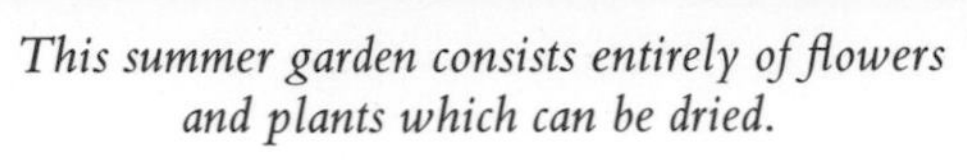

This summer garden consists entirely of flowers and plants which can be dried.

On the right is a perennial border, and on the left, at the back, are flower beds planted with annuals.

Eryngium planum

If you want really steely blue stems and heads, give the plant a very sunny, warm spot and a little extra lime in the soil.

Harvest when the whole plant is completely blue.

Flowering period: July–August.

Gypsophila paniculata (Baby's Breath)

This flower has become very fashionable in recent years. Almost every florist's bouquet is resplendent with Baby's Breath. Fresh bunches are sold the whole year round and are so attractive that they hardly need the addition of chrysanthemums or red carnations in my opinion; but tastes differ.

If you have it growing in your own garden, pick a large bunch, in full bloom, place it in a pot with a little water and leave it to dry out. It can also be hung up to dry.

Flowering period: July–August.

Lavandula augustifolia (Lavender)

Very well known as a fragrant herb but the blue flowers can also be used in dry bouquets to give a faint perfume.

Pick just before the flowers come into full bloom.

Flowering period: June–July.

Leontopodium alpinum (Edelweiss)

There is no need to go to the Alps to find them (don't pick them there); the plants grow equally well at home in a sunny spot in the rock garden.

Flowering period: July–August.

Liatris spicata (Blazing Star, Gay Feather)

This purple flower is very suitable for drying; the flowers bloom from the top of the spike to the bottom and should be picked when mostly open.

Flowering period: July–August.

Limonium latifolium (cultivated perennial Statice)

Blooms more extravagantly than the annual statice. The bluish flowers keep their colour well if picked in full bloom. The flowers do not open in wet weather; they can still be picked then but the colour tends to fade.

Flowering period: August.

Lythrum hybrid

Cultivated plants bloom in various shades of pink–purple. The wild Purple Loosestrife (*L. salicaria*) is also well known along the edges of streams. Flowering period: June–September.

They should be dried quickly to retain their beauty.

Monarda didyma (Bergamot)

A beautiful deep red flower with an unusual scent and an attractive inflorescence.

Pick when the tubular flowers have formed all the way round. Ideal for use in cheerful bouquets with grey-green shades, for example, combined with red and white roses.

Flowering period: July–August.

The leaves are delicious in a cup of tea.

The hybrid variety is also available in pink and purple.

Nepeta faassenii (Catmint)

Cats adore the smell of this plant while certain kinds of vermin (including lice) most definitely do not appreciate it.

Pale lilac-blue flowers, which should be picked when the spikes are for the most part open.

Flowering period: June–September.

Origanum vulgare (Wild Marjoram)

Pink flowers on long stems, which should be dried with the leaves on. When using in arrangements, keep some of the dark green leaves. It makes good filling material, and keeps its colour well. Any leaves which you remove can be dried for culinary use (see also page 80).

Flowering period: July until late autumn. The small flowers on the inflorescence do not open simultaneously. The time to pick is not so important; after drying, the flowers turn darker.

Some perennial plants:

1 Monkshood
2 Golden Rod
3 Achillea ptarmica (Sneezewort)
4 Anaphalis 'Schwefellicht' (Pearl Everlasting)
5 Anaphalis triplinervis (Pearl Everlasting)
6 Edelweiss
7 Acaena
8 Artemisia 'Silver Queen'
9 Astrantia alba and rubra
10 Santolina, not yet in bloom
11 Cultivated sea lavender
12 Centaurea macrocephalea

A Biedermeier bouquet (see page 63) which includes
1 Chinese Lanterns 3 Helichrysum
2 Hare's-tail 4 Fine Cloud Grass
5 Lonas inodora

Physalis franchetii (Chinese Lantern)

Everyone is familiar with this plant, arranged with honesty and sprays of oats in a leaky copper kettle – a breath of summer or eternal autumn? The red 'lanterns' assume a completely different appearance if, after drying, you cut the calyces into four to form star-shaped flowers, which should be mounted separately on wire.

Pick when the bell has completely changed colour (September).

Polygonum affine 'Darjeeling Red'

Evergreen ground cover, blooms with pink and dark red spikes. Pick when the spikes are fully in bloom.

Flowering period: August–September.

Santolina chamaecyparissus (Lavender Cotton)

A shrub-like plant which can be used for low hedges. The plant retains its attractive grey-white leaves, and blooms with lemon-yellow flower-buds.

Flowering period: June–July.

Solidago canadensis (Golden Rod)

An ideal flower for picking, with its deep yellow, fluffy panicles. The *canadensis* is the most popular species and is also found in the wild. Various hybrids have different shades (bright yellow and golden yellow) and panicles in different shapes. These need not be used whole; a few spikes can be tied together with wire and used in a modestly-sized bouquet, for example.

Pick in full bloom.

Flowering period: July–September.

Stachys lanata (Lamb's Tongue)

A low plant with grey, silky leaves, stems and spikes (the lilac inflorescence is of no importance for drying). The spikes continue growing for some time: picking should be timed accordingly. The leaves should be dried flat.

Lamb's Tongue picked before blooming is ideal for preservation, with leaves and all, in glycerine.

Callicarpa giraldiana (= C. bodinieri Giraldii)

A shrub which loses its leaves at the approach of winter but then displays the most beautiful purple berries. A larger harvest of berries can be obtained by placing several bushes close together to encourage pollination.

Pick before a hard frost.

Caryopteris × clandonensis

A very low shrub with sky-blue spikes.

Flowering period: September.

Beautiful branches can be cut in the autumn and hung to dry. The shrub blooms on one-year-old wood so that cutting the plant back in this way can only do it good.

When using in arrangements, tie the clusters of flowers together with wire.

Rhus cotinus (Cotinus coggyria) (Smoke tree)

A tall shrub, 2–3 metres (7–10ft) high, with attractive, rounded leaves. The 'Royal Purple' variety has deep red leaves. In July the shrub is completely covered in feathery flower stems, varying in colour from pink to red.

Picking time: August.

Rubus phoenicolasius (Wineberry)

Blooms with pale pink flowers. The bracts are of particular interest for drying. They are dark red, hairy and can be picked both before and after setting. The shrub also has attractively-shaped green leaves which are grey on the back and ideal for use in bouquet arrangements.

Varieties of hydrangea:
1 Pink Hydrangea macrophylla, dried
2 White Hydrangea macrophylla, dried
3 Blue Hydrangea macrophylla
4 Hydrangea petiolaris
5 Hydrangea paniculata
6 Hydrangea spec.

Hydrangeas and roses

You will no doubt have been hunting in vain through the foregoing lists of different varieties of flowers for the mainstays of flower-drying, hydrangeas (including hortensias) and roses – but they deserve a chapter of their own for the very reason that they are indispensable.

Hydrangea

This is a very extensive family of garden shrubs. I include here only the most common species: *Hydrangea macrophylla, Hydrangea petiolaris, Hydrangea paniculata.*

Hydrangea macrophylla (Hortensia)
An old-fashioned plant which can be grown both indoors and in the garden. It is unfortunately not very resistant to hard winters: the buds are formed in the autumn and next year's flowers run the risk of being frozen. This can be avoided by covering the plants with straw or packing it round them.

Cutting the flowers is good for the shrubs, but keep the stems short or you may accidentally cut off next year's buds.

Varieties of hydrangea found in the south-west of England and in Ireland are of a beauty and size not attained in colder areas. The colours there are overwhelming – great masses of intense blue and pink.

The colour, and particularly the way it changes in the autumn, is the most interesting factor. It can be influenced by burying old rusty nails, copper coins or alum in the ground near the roots (but I must confess that I have never been able to put this advice to the test).

When the flowers begin to bloom, they are a fairly even blue, white or pink. As the nights grow colder, the flowers change: they tend to become rather leathery to the touch and to turn slightly green; at the same time they may turn away from the light.

Flowers with mingled colours are particularly beautiful; matching colours is a simple matter, and sometimes the colours in a single bloom can form a guide for other flowers and seeds to be used in an arrangement.

Do not pick the flowers until they have turned colour and become slightly leathery to the touch. Look particularly at the centre of the bush; the most attractively coloured examples can often be found there. A hortensia which has not fully matured will become miserably shrivelled when picked.

Even when placed in water, hortensias in full bloom will only retain their beauty for a short time.

If the flowers are picked at the stage described above, they can be dried by many different methods: hanging, or flat if necessary, or standing upright in a vase in about 5cm (2ins) of water, allowing the water to evaporate slowly. If the petals start to wither when the water has dried up, a little more can be added. The only method of drying a young hortensia bloom is in sand. A large cluster of flowers can be divided into smaller bunches which can be placed in the sand, bunch by bunch, and dried as described on pages 17–21. Hortensias dry quickly by this method and retain their colour well.

The shape of the flowers makes them very useful in arrangements: they are excellent filling material. The flower-clusters are densely packed so that they form an ideal background.

Hydrangea petiolaris or Climbing Hortensia
This species is quite different from the common hydrangea. It is a climbing plant which grows very well on north walls and is self-supporting. It generally takes a couple of years before the plant really takes hold, but once it is growing well it soon covers a large area of wall.

The flowers, which appear on the shrub in June, have an attractive, lacy appearance due to a combination of small, fertile flowers in the centre and white sterile flowers on the outside. After some time the white flowers become green and the heads turn downwards; they are then ready for picking. While they are still white, they can only be dried in sand.

For a hanging decoration, for example, this is a very good flower to begin with: it is a good filler but nevertheless has style.

Hydrangea paniculata
This plant needs a rich, not too dry soil. The flowers consist of large, rather pear-shaped clusters of sterile florets.

At first they are white to pale pink, gradually fading to a combination of deep pink and green.

Hanging arrangement of roses and other flowers in shades of soft pink and green (for instructions, see page 62)

Pick them when the flowers have quite clearly faded and are papery to the touch. You can then simply hang them up, or stand them in a vase and leave the water to evaporate slowly.

This species is also ideal for arranging. The beautiful mingled colour combination is not too obtrusive in character and forms an excellent foundation (as in the illustration of the pink and green arrangement on this page). When dry the flowers do, however, become very brittle, so fine wire must be attached to the separate tufts.

Roses

The rose, in all its shapes and sizes, is well worth the trouble of drying, whether it is a wild species or the most carefully cultivated Tea Rose.

It is only in the last few years that roses have played a part in flower drying. They are usually used as rosebuds, but occasionally a beautiful, half-open rose finds a place in a bouquet. The simplest method is to dry buds or slightly opened flowers by hanging. A warm location is recommended since they retain their colour better if dried quickly. The flowers do shrivel a little and the colour turns slightly darker, but they are still recognizable as roses and the result is satisfying.

The most suitable roses for drying are those sold in the market in July and August for a relatively low price. Rosebuds from your garden can also be used. The prolific wild species will certainly not miss a few buds. Many of these species have a really lovely fragrance which is retained for a long period of time. The best result is obtained when all the petals of the bud

Flowers included in the arrangement opposite:
1 Quaking Grass 4 Polygonum affine (Knotweed)
2 Hydrangea paniculata 5 Love-in-a-Mist
3 Floribunda rose 6 Agrostis nebulosa (Cloud Grass)

can be seen. The first petals may already be curling.

The buds should usually be cut with fairly short stems, or too many immature buds will be cut off as well – but it is then difficult to tie them together. They can simply be spread out in a box in a dry, preferably warm, place and left to dry. If you have time, it is really best to mount the buds on wire stems immediately (see page 20), and make them into small bunches.

If you want a rose to remain really beautiful, you must use the sand-drying method (see page 17–21), even if you only preserve a couple of roses per year! Never use a rose which is fully open, or the petals will fall off immediately you remove the flower from the sand. It is, however, possible to dry an open rose if the petals are dried in the sand one by one and then glued back on to the receptacle, petal by petal. It is a very exacting, time-consuming job, but good results can be achieved. About four days is sufficient to dry a rose thoroughly in sand. A stem should be attached while the receptacle is still soft. It will later become extremely hard and a stem can then only be made by winding florist's wire around the original stem which can easily snap off, making it difficult to find another solution.

Roses mounted on stems can be made into bunches and hung up to dry. If you take care to bend the flowers away from each other, they will retain their shape perfectly. When dry, a little hair lacquer sprayed on as a protective layer will make the rose last longer. Do not overspray, or it will give the flower an unnatural shine. The best roses to use are those which have just begun to open. Many types of ordinary garden roses can be used, but cut roses from the florist are also ideal.

Single roses, many of the wild species for example, dry well in sand, but soon become limp if the air is too humid. Large, double roses, on the contrary, remain beautiful for years. A true red rose often turns very dark and the result may therefore be rather disappointing. Since most colours turn darker when dried in sand, lighter colours are preferable.

After the buds and flowers, you may want to preserve the rose hips, but most varieties of hips shrivel so much that they lose a great part of their charm. The smaller the hips, the better the results. The most suitable are the hips of the *Rosa multiflora.*

In spite of many attempts, I have never succeeded in finding a satisfactory method of preserving berries. Any suggestions will be more than welcome.

A collection of summer wild flowers
which can be found on banks and other rough ground.
The following can be dried:

1 Tansy
2 Mugwort
3 Varieties of grass
4 Shepherd's Purse
5 Poppy (the seed-heads can be dried later)
6 Sorrel

Wild flowers

Picking wild flowers is something which almost everyone finds irresistible from time to time. Children pick bouquets of wild flowers and make daisies into chains; and people setting off for a walk not infrequently return with a bunch of flowers picked by the wayside.

This is a harmless pastime, but we are all becoming aware of the fact that the variety of plant-life in our natural environment is increasingly threatened. It is therefore important to use common sense and consideration when picking.

There are, however, a few rules to remember. At the beginning of June, for example, you can with a clear conscience and without causing damage, pick plenty of red sorrel, white yarrow, and especially wild grasses of which there are very many species. Always use a sharp knife to cut the stems, otherwise you may easily pull whole clumps out of the ground. Wild plants should never be uprooted. And never pick all the flowers in a clump; if necessary, go even a mile further to find the next one.

In recent years more concern has been shown for the environment in the maintenance of roadside verges. In general they are now mowed twice a year, in early summer and in autumn; and there is so much protest against the spraying of edges and verges with all kinds of weedkiller that this, too, is done increasingly less often.

Now a few practical tips.

Although the weather can cause problems, it is better to pick flowers only in dry and preferably sunny weather. This can be ignored in urgent cases, but your harvest must be dried quickly or mould will definitely start to form. The best place to hang your flowers and leaves is in a boiler room. The central heating can, of course, be turned on, but that is becoming an expensive business.

In some countries, various species of wild flowers of increasing rarity are now protected by law and must not be picked. You can, and should, obtain information about this from the local authorities. The following list gives details of a selection of reasonably common wild flowers which are good for drying. This is only a limited list, but you can experiment with other varieties or refer to a book on wild flowers for further information.

Achillea millefolium (Yarrow)

The name of this plant comes from the Greek hero, Achilles, who used the herb to heal his friends' wounds.

You may pick as much of it as you like.

The large white umbels can be seen along pathways, roadside verges, even on the central verge of major roads, and on fallow land.

Always take a sharp knife with you. The stems are very tough and when they are picked by hand the whole plant is often pulled out.

Yarrow can be found in June and at the beginning of July but if you have no success with the first picking, you

can gather fresh stores from the second flowering. Pick at just the right moment, when the centres of the small white florets which form the umbel are clearly visible – otherwise they will shrivel up when dried.

The attractive, dark brown umbels of flowers which have finished blooming can also be used. They can often be found right up to early winter.

White achillea is indispensable as a background for bouquets.

Alisma plantago-aquatica (Common Water Plantain)

Since less poison is now being sprayed in the course of maintaining waterways, the common water plantain can now more often be found beside rivers, canals and ditches.

The small white or pink flowers bloom throughout the summer, but here we are only concerned with the seeds. The plant forms a very decorative, small Christmas-tree shape with seed capsules on the branch terminals which can be found in ditches and beside slow-flowing rivers, canals, ponds etc. from September onwards.

At this stage, drying is not necessary. Take care that the plants do not become entangled: hang or stand them separately from one another.

The Water Plantain is ideal for use in Christmas decorations (see page 77).

Arctium lappa (Greater Burdock)

This plant does not originate, as the Latin name would seem to suggest, from the polar regions.

These are the large burrs which provide schoolchildren with so much fun when stuck in someone else's hair. They have a more friendly application in dried-flower bouquets and – if sprayed with a dash of gold paint – in Christmas decorations.

The greater burdock can be found from August onwards on roadside verges and rough places, especially near large rivers.

It can be picked green, but can also be used when it has turned brown.

You can perhaps make use of the adhering property of the burrs by simply sticking them on your arrangement without using artificial stems.

Armeria maritima (Sea pink, Thrift)

Thrift is a well-known plant with beautiful silver-pink flowers, commonly found along the coast. It is not (yet) protected, but it should be picked with care as it is easy to pull up a whole clump by mistake. Pick the flowers before they reach full bloom; the flowering period extends from early May through to September.

Hang up to dry.

Cultivated varieties are also available (see page 30).

Artemisia vulgaris (Mugwort)

This tall, grey-green plant is commonly found on rough ground and demolition sites. It often provides some visual relief in cheerless areas of urban development.

The rather insignificant but abundant greyish florets form attractive grey spikes. Both their shape and their colour make them indispensable in bouquets. One word of warning: mice are crazy about them! Small heaps of grey pellets beneath the hanging bunches immediately reveal the presence of these unwanted guests.

The flowering period runs from July to September.

Hang up to dry.

There are also many cultivated species of *artemisia* (see page 31).

Calluna (Heather)

If you do it as carefully as the sheep, you may pick a little heather. Pick it preferably when in full bloom. Before using heather in bouquets, the flowers should be sprayed with a little cheap hair lacquer, since heather soon disintegrates when very dry. (See also the glycerine method, page 17.)

Picking time is August–September.

A piece of white heather is a symbol of good luck: a small dried bouquet which includes white heather and a few rose-buds could make a gift for someone special.

Carex (Sedge)

Many kinds of sedge (and there are dozens of different species) stand with their feet in water, making picking rather difficult. You will need a knife to cut the stems.

There are many variations of inflorescence, but it is always brown.

Pick before fully mature.

Carlina vulgaris (Carline Thistle)

Pale white receptacle surrounded by silver bracts. Suitable for large arrangements, but handling it can be a painful business.

Flowering period: July–September.

Chrysanthemum vulgare (=Tanacetum vulgare) (Tansy)

Children call the flowers buttons. You can find them from July to October on dry verges and sandy wasteland. They should be picked when the buttons are bright yellow and round.

Cirsium vulgare (Spear Thistle)

One of the many species of thistle found in meadows and verges, often to the considerable regret of farmers.

Most varieties bloom from July to August.

If you pick the thistles when the first purple petals appear, you can count on having purple flowers throughout the winter; but if you pick them in full bloom, you will end up with large handfuls of grey down. No need for despair, however: if you remove the down, you will be left with beautiful, glossy receptacles which are excellent for use in arrangements. It is possible to see in the receptacle the regular pattern formed by the down.

Daucus carota (Wild Carrot)

Wild carrot is a well-known plant, found particularly on roadside verges. The blooms are white umbels, the edges of which, immediately after flowering, curl over to form a kind of green, lacy ball. If these balls are picked when green, they are very decorative in all kinds of arrangement. You can expect the flowers to be at this stage from August onwards.

Filipendula ulmaria (Meadowsweet)

The beautiful white crests of meadowsweet grow in ditches, alongside canals and in wet meadowland.

For drying purposes it is the seeds, which grow in clusters of bright green, spiral-shaped capsules, which are interesting.

In September they will still be a lovely green colour. Later they will turn brown and disintegrate easily.

The attractively shaped seed-heads of meadowsweet are indispensable in large bouquets.

Heracleum sphondylium (Hogweed, Cow Parsnip)

The hogweed's large seed umbel can be used for decoration in innumerable ways.

This is the common form of hogweed. There are many other varieties, some of which can grow to the height of a man, or more. The flowers often have a slightly pink tint. They should be picked for drying when the seeds have just developed, forming attractive green rosettes which retain their colour for a long period.

But take care! Some people are allergic to this plant and if it touches the skin it can cause a painful rash which is sometimes visible for months.

Humulus (Hop)

Wild hop climbs with long tendrils in hedges and undergrowth, on moist ground. Hop can also be grown in your own garden from seed, but it requires plenty of space. Apart from the usual plant, there is also a variegated variety of hop.

The beautiful green bracts of the female cone-like spikes unfortunately soon turn yellow in the sunlight and nothing can be done about it.

Dry by hanging.

Hogweed (Cow Parsnip) in full bloom.
To dry this plant you must wait until
the seeds have formed.

Juncus articulatus (Jointed Rush)

A very decorative marsh plant, with umbels covered with glossy, dark brown fruit capsules.

Jointed Rushes can be found in July and August in marshy places. It takes some time to collect a reasonable bunch, but it is worth the trouble: after hanging for a year, they still look as though you picked them yesterday.

Lapsana communis (Nipplewort)

If you find this in the garden when you are weeding, leave a few growing. The delicate capitula of tiny achenes (fruit) are ideal for arrangements. This is a common plant in the wild and can be found without difficulty throughout the summer.

Limonium vulgare (Sea Lavender)

This plant, though relatively rare, is often abundant in coastal areas (salt marshes). Where growth is prolific, it can produce a sea of blue and it is tempting to snip off a flower here and there. But you should resist the temptation. Similar-looking plants – often known as Statice – can be easily cultivated (see page 34).

Another solution is to bring bunches home when you go on holiday to areas where it is not a threatened species.

Pick when the flowers are wide open.

The blooming period is from July up to and including October.

Lythrum salicaria (Purple Loosestrife, Horsetail)

Purple Loosestrife grows in ditches and marshes or on riverbanks. The long spikes are deep purple. This species has also suffered from the spraying of ditches and verges with chemicals, but it is now becoming more common again.

Care should still be taken with picking, however. The flowers should be picked in full bloom.

It is essential to dry them quickly in order to preserve the colour.

They can be found from June to the end of August.

Mentha aquatica (Water Mint)

You can often smell water mint before you see it.

Large quantities of this pale purple flower can often be found on marshy ground and alongside ditches.

If picked in full bloom, the flowers retain their colour well. The aroma preserves even better, so if the mint is not used in a bouquet, it can always be made into mint tea.

The flowering period is from June to the end of September.

Rumex (Sorrel)

The numerous species of sorrel can be found on all types of soil, and there is considerable variation in height.

The colours vary from pale green to dark brown. All varieties can be dried.

The flowering period is May – November. Pick preferably in full

1 Plantain 4 Salvia
2 Dead nettle 5 Nipplewort
3 Meadowsweet 6 Burdock
7 Tansy

bloom. If this is impossible for some reason, sorrel can also be picked after the formation of seed, but it will then disintegrate more easily when used for winter. Hair lacquer can provide a solution here too.

Sanguisorba officinalis (Great Burnet)

The Great Burnet is locally common, found in wet grassland. The compact, dark red spikes can be successfully and easily used in arrangements. Perhaps you have a wild flower garden where it grows. Pick at the beginning of the flowering period. The flowers fall apart easily, so it is best to spray them with hair lacquer.

The flowering period is July.

Solidago virgaurea (Golden Rod)

Golden Rod can sometimes still be found wild in large quantities on sandy ground and dry banks – alongside railway tracks, for example. Many species of solidago can, of course, also be obtained from nurseries for the garden (see page 37). Pick the flowers just before full bloom. If you pick them too late they will turn downy.

The flowering period is end of July and throughout August.

Trifolium arvense (Hare's-foot)

Not to be confused with hare's-tail, which is a species of grass. Hare's-foot is the soft, fluffy, grey-pink cylindrical head very commonly found on sandy verges. It can be picked throughout the summer.

The heads are very attractive in bouquets but should be well sprayed with hair lacquer to prevent disintegration.

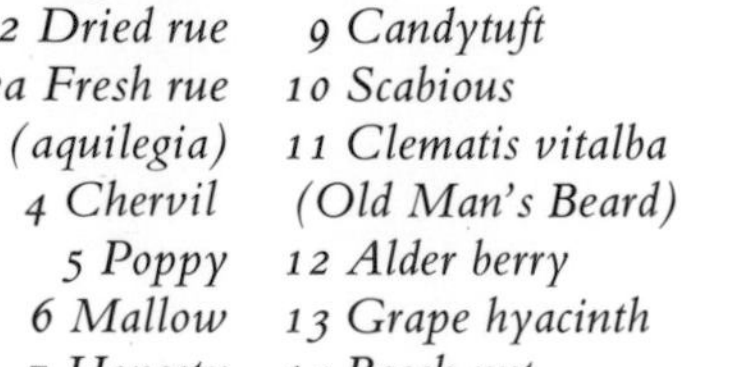

Seed-heads and seed-pods

1 Campanula
2 Dried rue
2a Fresh rue
3 Columbine (aquilegia)
4 Chervil
5 Poppy
6 Mallow
7 Honesty
8 Dill
9 Candytuft
10 Scabious
11 Clematis vitalba (Old Man's Beard)
12 Alder berry
13 Grape hyacinth
14 Beech nut

Grasses and seeds

Many grasses and seeds could be included in any of the plant groups already discussed. Grasses form a very extensive family which includes more than two hundred species and all of them can be dried with varying degrees of success. Grasses are used in many arrangements.

There is an even greater variety, if possible, of seed-heads and seed pods. Not all, unfortunately, are suitable for drying or arranging. It is worth taking a good look at the different fruit settings of flowers. You will be amazed at the beautiful shapes and the great variety. Since the colours are generally unobtrusive (with the obvious exception of red berries), seeds or fruits are chosen principally for their shape.

There is space here to mention only a few of the attractive grasses suitable for drying and the various kinds of seed-heads with well-defined shapes. You will be able to find many more species for drying and arranging to suit your own taste and ideas.

A good general rule for grasses is to pick them just before they flower; but many varieties remain attractive and will keep well after blooming.

Grasses or corn with heavy grain should be picked immediately after the fruit sets; they should still be green. Otherwise the ripe grain will often fall off the ear, or the ear will disintegrate.

Most grasses are dried by hanging, tied in bunches. The more delicate grasses are better dried upright (see page 13).

To avoid breaking the fine stems, wrap raffia or florist's wire spirally round the bunches.

Many seed-heads can be picked while still greenish. If you wait longer, they will turn brown and may soon fall off. This is not necessarily a drawback, as each stage has its own particular charm, so it is a good idea to pick some seeds early and some late. Feathery seed-heads such as *Clematis tangutica* and the Pasque Flower (*Anemone pulsatilla*) must be picked as early as possible and immediately sprayed with hair lacquer if necessary.

Cultivated annual species of grass

Agrostis nebulosa (Cloud Grass)
Briza maxima (Great Quaking Grass, or Pearl Grass)
Briza minima (Small Quaking Grass)
Hordeum jubatum (Squirrel-tail Grass)
Lagurus ovatus (Hare's-tail Grass)
Phalaris canariensis (Canary Grass)

These species of grass are grown as annuals; sow at the end of April in a sunny spot.

Many of these grasses propagate themselves quite easily. If you don't weed or hoe too vigorously, there is a good chance that they will come up again the following summer.

Cultivated perennial species of grass

Cortaderia argenteum (Pampas Grass)
Pennisetum japonicum
Stipa pennata (Feather Grass)

Some species of wild grass

Bromus sterilis (Barren Brome Grass)
Festuca pratensis (Meadow Fescue Grass)
Glyceria maxima (Reed Sweet Grass)
Holcus lanatus (Yorkshire Fog)
Hordeum murinum (Wild Barley)
Phragmitis australis (= P. communis) (Common Reed)
Poa pratensis (Common Meadow Grass)

As already mentioned, this is only a small selection. For a detailed list and for visual identification, consult a good wild flower book.

Seed-heads and seed pods suitable for drying

Alnus glutinosa (Common Alder)
The alder can be found in damp woods and beside streams and ditches. The female berries are green in autumn and a grey-brown colour in early spring.

Aquilegia (Columbine)
The seed pod is an attractive shape, with petal-shaped sections.

Clematis tangutica
Pick immediately after blooming, before the styles turn feathery; spray with hair lacquer before use.

Little material is needed for a simple arrangement. Several kinds of grass, Lady's Mantle, and the seed-heads of dill and Star Anise are used here.

Euonymus (Common Spindle Tree)
The pink hood-like fruits tend to shrink a little but the orange berries preserve well.

Fagus sylvatica (Common Beech)
The inside of the spiny seed case of the beech nut stays a beautiful brown colour.

Iberis (Candytuft)
The seed pods are winged and notched.

Larix (Larch)
Larch cones can be mounted individually on wire and used in decorations.

Lunaria (Honesty)
After drying, remove the outer layers of the seed pods to expose the transparent, silvery inner lining. If the seed heads are picked green, the pods, when peeled, will be a greenish colour.

Malva (Mallow)
Beautiful green fruit. When picked late, the seed pods are grey.

Muscari (Grape Hyacinth)
Pick when the seed pods burst open.

Papaver (Poppy)
Both wild and cultivated species can be used.

Ruta graveolens (Rue)

Scabiosa caucasica (Scabious)
Pick when the petals have faded to leave a beautiful grey-green flowerhead.

Arranging

The time will come when you suddenly realize that the whole house is full of dried flowers and foliage. Leaves in newspaper lie under the carpet and couch, and containers and pots fill every nook and cranny; even the ceilings are here and there a sea of colour. At this point you will undoubtedly begin to wonder what you are going to do with them all.

If you don't need the whole of the season's harvest, your friends may, of course, appreciate being presented with bunches of dried flowers; but you will soon find that using them all for your own arrangements is even more satisfactory. It is only possible here to give a few instructions, ideas and tips for the best ways to use your material. The rest you must discover for yourself, using your own creativity and ingenuity.

Each person's collection will be different. It will usually be determined by the contents of your own garden, the surrounding country, or of material brought back from holidays. The style and character of your house and your own personal taste will also contribute to the way in which you decide to use your flowers.

Before individual flower arrangements are discussed, here are a few general ideas, all of which are simple to carry out.

Bunches hung up to dry are very decorative in themselves. How and where you hang them depends on the type of house you live in.

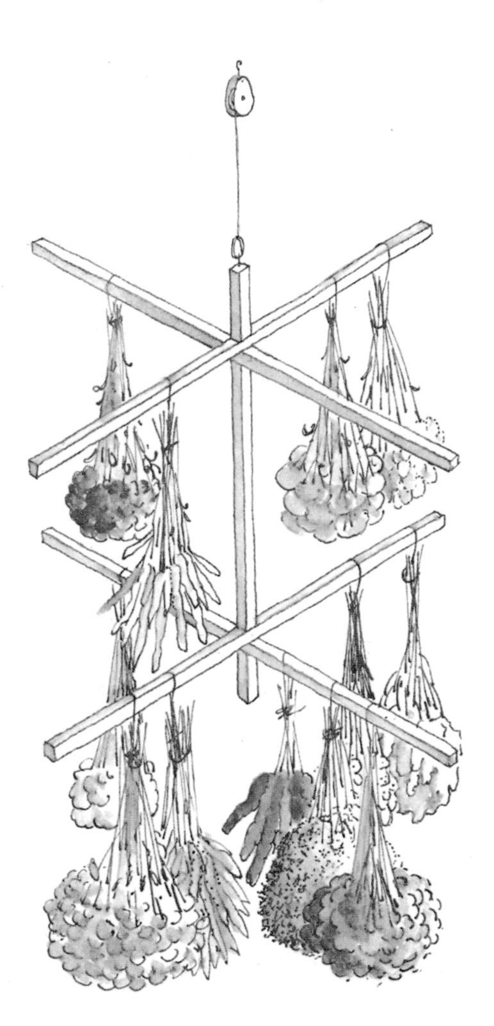

Good results can be obtained by using a flower-drying rack, hung at the required height by means of a pulley, as shown in the illustration. A broom handle can be used in the same way, hung on two pulleys (see next page).

Another, very simple, solution is to put up a length of clothes line and hang the bunches on it with small clothes pegs. It is possible in this way to disguise all kinds of imperfections in the home. High ceilings are less noticeable, the back of a staircase takes on a more cheerful aspect and dark corners in an attic room become less dull. Even the most unattractive passage is considerably improved.

Subtlety is needed in the use of colours. Combinations of blue, white and grey, white and yellow, or a gay mixture of flowers in graded colours can be very attractive.

A few good sprays of hop with plenty of bracts or 'bells' can be strung along a door post or in a narrow passage between two rooms. They can be hung in place freshly picked, and left to dry. Since the bracts are the most interesting part, the leaves can be cut off, but this is not essential.

Using flowers in bunches

Composing a dry bouquet can be a complicated and often time-consuming business, but this need not be the case if you use bunches of different kinds of flowers. Large baskets or coarse earthenware pots are ideal containers.

Begin by placing bunches around the rim of the pot or basket. The stems of these bunches need not be long. When you have completed the first ring round the edge, start the next. As you progress towards the centre it is better to leave the stems longer to build up the height and give unity to the whole. Aids such as Oasis or florist's wire are unnecessary if the arrangement is closely packed.

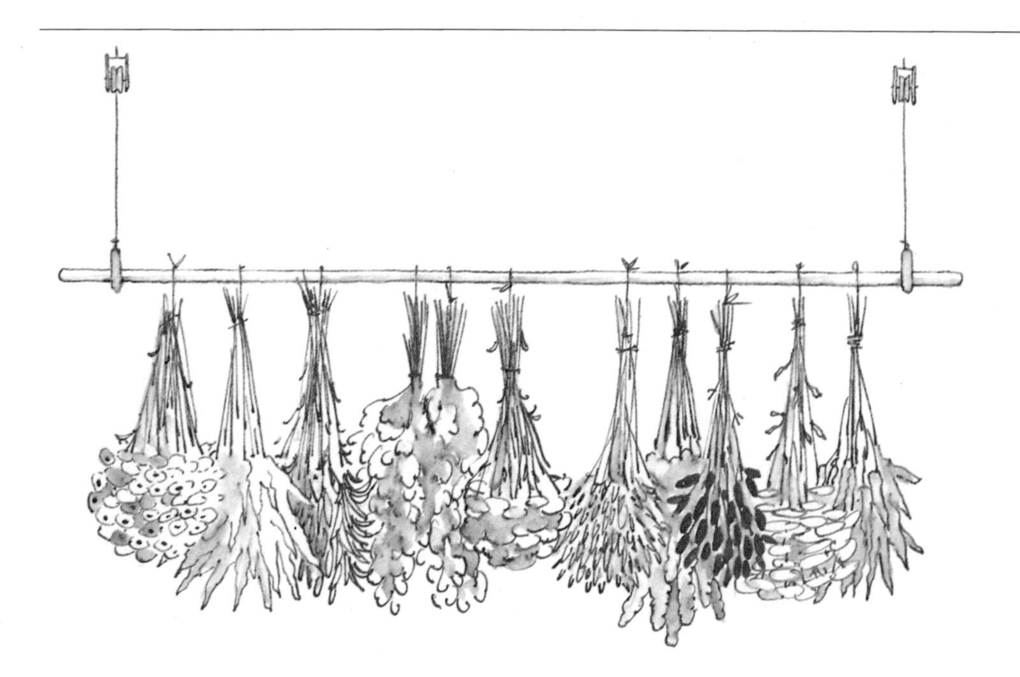

This kind of arrangement is always very effective, especially if a mixture of bright colours is used.

Scent baskets or pots

A scent basket is a kind of instant *pot pourri*. It exudes a delicious fragrance which, surprisingly enough, lingers for years. Making a good *pot pourri* is an art which requires a certain amount of expertise. This is an attractive substitute.

When making dry bouquets, you will usually be left with a large number of flower-heads which have fallen off. Collect them together and keep them. Put some of these flowers in an attractive pot or basket, with or without a lid. Using leftovers will give you a mixture of colours.

Alternatively, you can remove the stems from a selection of flowers at the end of the season, choosing any colour combination that suits the pot or the decor.

Sprinkle a few cotton-wool balls, in matching colours if possible, with an aromatic oil (obtainable from specialist suppliers) and hide them among the flowers – which you may or may not have arranged in regular patterns. Finally add a layer of flowers placed very carefully with the heads facing upwards. Should the perfume fade after a while, the cotton wool can be resprayed and tucked between the flowers again.

You can, of course, match the flowers to the perfume you use. A lavender or rose fragrance can be heightened by the use of real lavender flowers or rosebuds.

An attractive scent pot of this kind is a delight in any part of the house – living-room, guest-room or bathroom, for example – and makes a much appreciated gift.

Decorating a Mirror

An edging of dried flowers immediately brightens up an ordinary mirror, and is very simple to make.

All you need is an unframed mirror, a little transparent glue and some dried flowers, mosses, etc.

Decide beforehand which flowers, leaves and mosses you are going to use and place them ready. Now smear a few inches along the edge of the mirror with glue and do the same to the flowers and other material you want to place on those few inches.

Place the flowers carefully in position and press for a moment to give them time to adhere firmly, so that they will stay in place if you turn the mirror over. Then continue with the next section till the 'frame' is complete.

A wall decoration can be made on a piece of wire mesh. This arrangement includes yellow anaphalis, orange and yellow helichrysum, Lady's Mantle, white acroclinium, several varieties of grass, and moss.

On the mirror illustrated, I first arranged a bed of moss, to which small sprigs of golden rod, bright yellow flower-heads of *Helipterum sanfordii* and an occasional yellow rosebud were then glued. The mirror's oval shape was accentuated by adding a longer arrangement at the bottom, consisting of a few ferns and the same flowers as were used around the edges.

The bed of moss is not essential; the flowers can be glued directly to the mirror.

Arrangements of dried flowers

Flowers can be arranged in an infinite number of ways. All kinds of styles can be learnt at flower-arranging classes, which are given everywhere nowadays. The art of arranging flowers is not very old in the western world – in fact it dates from about 1800. Before that time it was mainly an occupation for artists, who composed on canvas exquisite arrangements which must have been painted from imagination rather than from life, since they usually consisted of flowers which bloom at different seasons. From the beginning of the nineteenth century, pot plants and flower arrangements are increasingly evident in illustrations of interiors. The decorative use of both fresh and dried flowers is especially notable in the Biedermeier period.

The romantic, usually circular, Biedermeier bouquet, composed of many varieties of flowers, is still familiar today.

Dried flower arrangements were

formerly often placed under glass domes to protect them from dust. In this way they could be kept for a very long time. You can sometimes still find one of these old, faded bouquets in an attic or junk shop.

During the nineteenth century, much was also learnt in Europe about the art of Japanese flower arrangement, ikebana. The art is still highly esteemed in oriental countries – and in the west too. These arrangements often have a symbolic meaning; and though usually made up of only a few twigs or flowers they can suggest an entire landscape.

But although a single branch of blossom in a vase can be very beautiful, unless you have studied the art and symbolism of ikebana it is probably better to stick to traditional western methods of using flowers.

In dried flower arrangements, as in paintings, flowers from all seasons can be used together.

Since dried flowers do not need to stand in water, the possibilities are numerous: wreaths, garlands, entire wall decorations and, of course, bouquets, to name only a few.

Various aids can be used to hold the flowers in position. Oasis, the rigid green foam into which the stems can be pushed, has made arranging very much easier. Kneadable material (mastic), wire and mesh are also employed. A list of materials, with illustrations, can be found on page 92.

When choosing plant material for an arrangement, it is important to bear in mind both colour and shape.

Colour selection is possibly the most personal element in flower arrangement and few directives can be given. A preference for one colour or combination of colours will vary with each

Included in the wreath opposite:
1 Wild carrot 3 Yarrow
2 Statice 4 Allium
5 Mugwort

individual. There are some people who only like yellow-orange arrangements and others who prefer blue.

As dried flowers age, the colour tends to change and the green of the stems and leaves will turn rather yellow. This should be borne in mind when you select your colours.

A pink bouquet which includes a lot of green loses its green freshness after half a year and the combination of drab yellow with pink is not attractive.

Blue and grey, on the other hand, are good lasting colours and an arrangement composed of them will retain its character for a long time.

The second important point is shape. A bouquet consisting entirely of spherical flower-heads and/or seeds would make an extremely dull design; and since colours lose their clarity in the long run, an attractive silhouette is important to keep the arrangement lively.

A major problem is that working with dried flowers creates an incredible mess. Broken-off flower-heads, stem remains and pieces of Oasis are mixed up with lethal oddments of wire, and bunches of headless stems tend to pile up high. In addition, you will often only use half a bunch when making an arrangement so the other half will be left lying on your work-table too.

It is best to have a special corner where the mess can be ignored for a while. It's amazing what attractive compositions can later be achieved from that disorderly pile of bits and pieces.

Arranging in wire mesh

Whole areas of wall can be covered with flowers in this way – in a dark passage, for example, or for an area above a door, or on a narrow wall between two windows.

Take a piece of coarse wire mesh, plastic-coated if possible, measuring a little larger than the area to be covered so that when it is fixed to the wall by means of a few nails it will bulge slightly. Large or small bunches of flowers, grasses and moss can now be worked into the mesh, with or without a fixed pattern.

If the bunches seem likely to fall out, you can tie them to the mesh with green florist's wire.

The arrangement illustrated is composed of many shades of yellow and white, enlivened by the use of grass and moss. One of the children has even stuffed an old bird's nest into the middle.

A combination of single colours is one possibility, but there are, of course, many others: you could even reproduce a plan of your own garden. In any case, you will need a large quantity of flowers. Bunches left over at the end of a season are ideal here.

If later you want to replace a few bunches in the flower-wall, there is no

A flower wreath

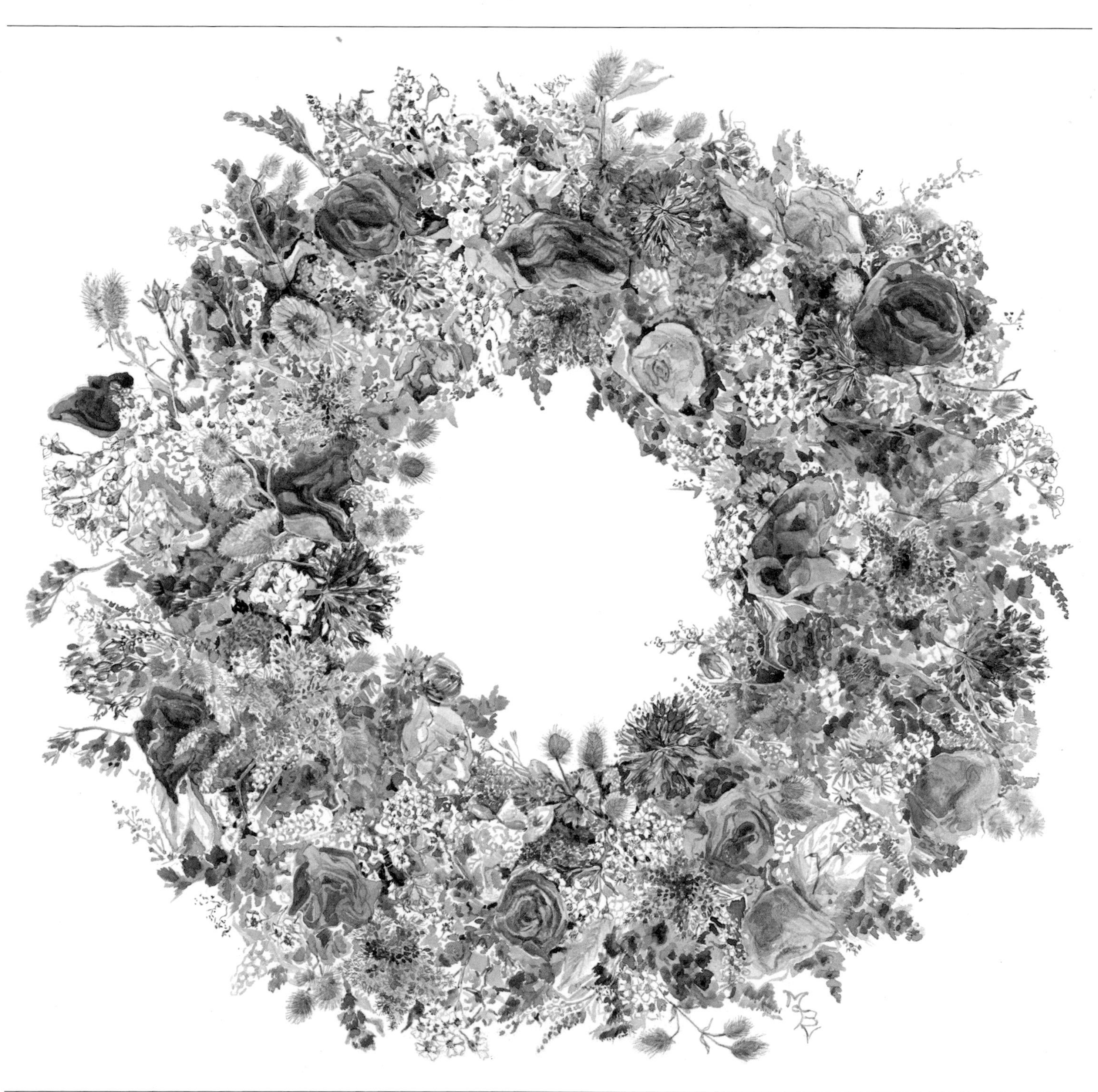

A hanging bouquet. An oval-shaped piece of Oasis was used as a base (see pages 62 and 93). Hortensias and anaphalis were used as filling material, and the brown beech leaves give the arrangement depth.

problem: simply remove the old bunches from the mesh and insert new ones in the spaces. In this way the wall will continually look new and fresh.

Wreaths and other shapes

A different technique is used for making wreaths, hearts and horseshoes. Formerly, clay or rolls of wire mesh had to be used; but now Oasis, a much easier material to work with, can be used as a base. The required shape can be cut out of a brick of Oasis. (You will probably have to ask your florist to order large bricks for you.)

The base for a wreath is made as follows: place a plate or dish of the right size on a flat brick of Oasis and cut out a circle round it with a sharp knife. Then place a smaller dish on this circle and cut out a second circle (see page 93). Trim the upper vertical edges to give the wreath a rounded top.

You now have the base for a wreath

The flowers below were also included in the arrangement shown opposite:
1 Delphinium 'Pacific Giant' 3 Achillea filipendulina
2 Sorrel 4 Yellow roses

and also a circle from which additional smaller wreaths can be cut, using small bowls to cut round, for example, thus producing a series of wreaths in varying sizes.

A wreath of flowers is always appealing and evokes a somewhat romantic atmosphere.

Since it is based on a flat surface, there are no problems with the construction or the proportions of the wreath as a whole. It is best to cover the bare Oasis first with reindeer moss. If you cannot gather this moss in the wild, it will have to be ordered from the florist. Alternatively, you can use green sphagnum moss. First soak the moss in a little luke-warm water; this makes it supple and easier to mould. Squeeze it out well and fix it to the Oasis with special staples or pieces of wire bent to the shape of a hairpin.

Now fix each flower onto the wreath, one kind after the other, ensuring that they are evenly distributed over the whole surface. Tie very small flowers together in bunches with wire. Larger flowers can simply be pricked into the Oasis with their own stems or on artificial stems. It is best to start with flowers which cover the background well; for example, achillea, hortensia, anaphalis or spirea. These can be followed by the more showy flower-heads such as xeranthemum, globe amaranth, chives or helichrysum.

Since the arrangement will now probably look rather stiff, the contours can be softened with spike-shaped flowers or grasses; suitable kinds are

larkspur, *statice suworowii*, Lamb's Tongue, varieties of artemisia; wild hare's-foot is also good for this purpose.

Lastly, a place can be found for a few beautiful roses, zinnias or hollyhocks. If these are added at the final stage you will not need many of them and you won't be tempted to fill empty spaces at random with these special flowers; they should be placed where they can be seen to their best advantage.

The finishing touches can be added with fine grasses or Baby's Breath. Make sure that the wreath is well filled all round.

Hanging arrangements

These can be made in either a round, oval or oblong shape. Again, it is no longer necessary to use crumpled chicken wire as a base. Special equipment now available makes the work considerably easier.

So-called 'bouquet holders' are available in the trade in various sizes and can be ordered from florists. They are bricks of Oasis in a plastic holder with a hanger (illustrated on page 93). As they are rectangular, an arrangement based on them will be roughly oval. If you want a different shape you can use bricks of Oasis cut to shape, as for a wreath.

For very long or large arrangements two or more bricks of Oasis can be fixed together by running strong wires through them (see illustration, page 92). When the wires have pierced through the bottom brick, bend the points into hooks and then pull the wires back up until the hooks grip the Oasis firmly. The wires protruding at the top can be bent into loops for hanging up the arrangement.

A small round base can be cut out with the help of a saucer. The vertical edges should be well smoothed off.

A hanger can easily be made on this shape too, by pushing a long piece of strong wire bent double through the Oasis, bending it into hooks at the bottom and pulling the wire back until it grips the Oasis, leaving a loop at the top.

Now that the technical problems have been solved, you can begin by covering the base with moist reindeer or sphagnum moss.

If you want your arrangement to have a shape different from the base, start by defining the outer contours by means of long flower spikes, seed-heads and branches with leaves, or ferns, and then work within this outline. It is possible in this way to make a round, oval or oblong arrangement based on a rectangular bouquet holder (see illustration on page 93).

Flowers that make good fillers should be used next – for example, *Hydrangea paniculata* or one of the many umbelliferous flowers such as achillea, anaphalis and tansy. More delicate material such as meadowsweet, tarragon, Lady's Mantle and golden rod is ideal for filling up large gaps.

The work sequence for the remaining material you intend to use is the same as for the wreath. Start with the distinctive flowers (helichrysum, xeranthemum, acroclinium, globe thistles and rhodanthe tied in bunches, for example) and then add some spiked flowers (larkspur, polygonum, artemisia 'Silver Queen' or mugwort) to enliven the shape.

Grasses and Baby's Breath come last to soften the contours and fill up any gaps.

If a bouquet holder is not used, the arrangement may resemble a rather flat pancake. To avoid this it is advisable, right from the start, to leave the stems of the flowers in the middle a little longer so that they stand out. If you start off with a good shape, it is easier to keep to it, but it is difficult to correct a bad shape while the work is in progress.

For those with no experience, it is simplest to begin symmetrically: for example, a larkspur placed top left can be balanced by another larkspur top right. After a time, when you no longer find it a problem to keep the composition round or oval, this system can be dropped.

Vases

It can be difficult to find just the right vase for fresh flowers, but with dried flowers there is no problem. Literally any kind of holder can be used, from a wooden egg-cup to a Wedgwood vase.

Here are a few suggestions: antique rose holders, sherry glasses, copper kettles, ash-trays, gravy boats, candle-sticks, hanging salt pots, etc., etc. Baskets in all shapes and sizes are particularly good as their natural material goes well with dried flowers.

There are nevertheless a number of points to be borne in mind. Do not use a highly ornate container whose colours would continually have to be taken into account. Vases with a small neck are not really to be recommended either, as it is difficult to insert the Oasis securely enough. But with a little ingenuity, much can be achieved.

It is absolutely essential for the Oasis to fit tightly. There is nothing more annoying than an arrangement which falls out of the vase at the slightest knock. It is now possible to obtain plastic prongs which can be glued to the bottom of the container, on which the Oasis can then be spiked. Do not try to glue the Oasis itself to the container as it will not take either adhesive or sellotape.

To cut a piece of foam for a vase to the correct size, press the opening of the vase on to the Oasis and then try to cut out the exact shape with a sharp knife. If the piece of foam turns out to be too small for the vase, wedge it with pieces of the same material.

Vases or glasses on a base or stem are very suitable since the leaves, seeds and grasses have room to hang attractively over the edge. To achieve this effect, you must allow the Oasis to stand an inch or two above the rim so that overhanging branches or flowers can be inserted around the sides.

For a triangular bouquet (i.e. with a flat back – see page 65), it is a good idea to leave a fairly large amount of Oasis above the rim. It is then much easier to achieve the required height of the bouquet (which should be one and a half times that of the vase) and fewer flowers will be needed for the 'heart' of the arrangement.

To prevent the Oasis showing through a glass container, moss or leaves the colour of the bouquet can be placed between the foam and the glass.

When using a basket, you can press a lump of ordinary modelling clay on the bottom, to which the Oasis can be fixed by means of long wires. They should be inserted right through the Oasis and as far as possible into the clay. The clay makes the light basket heavier and more stable and it shrinks as it dries, so that the wires are held firmly in place.

A Biedermeier bouquet

When making a Biedermeier bouquet, the order of work is the same as for a hanging arrangement. The only controlling factor is the shape of the vase.

The bouquet must grow out of the vase, as it were, and not lie on top like a flat lid.

For the sake of clarity, the order of work, once again point by point, is as follows:

1 Fix the Oasis securely and cover with moss. Reindeer moss must first be soaked in luke-warm water to make it supple. The green Oasis is concealed by the moss and holds the flower stems firmly in place, making the whole arrangement more secure.

2 Now add the filling materials such as hortensias and all kinds of umbelliferous flowers – meadowsweet and statice, for example.

3 Outline the shape of the bouquet with tall flowers, seed-heads or leaves; for example, larkspur, polygonum, mugwort, honesty and ferns.

4 Now add the more distinctive flowers such as helichrysum, acroclinium, thistles and love-in-a-mist. Smaller flowers such as rhodanthe or xeranthemum and ammobium are best tied in small bunches; a single flower of these varieties would not be effective in a larger type of bouquet.

A triangular (flat-backed) bouquet in shades of blue and grey. Hortensias, blue and white statice and anaphalis make up the background. The feathery seed-heads of the Pasque Flower (Anemone pulsatilla) give a lighter effect.

Also included in the arrangement opposite:
1 Hare's-foot 3 Globe thistle
2 Lamb's Tongue 4 Scabious
5 Carline thistle

5 Last of all add the special flowers. It is important to arrange them so that they do not upset the uniformity usually aimed for in a round bouquet of this kind. This generally means using an uneven number of flowers: an even number is inclined to divide the arrangement into sections, destroying the unity of the composition.

6 The addition of some Baby's Breath, pearl grass or Lady's Mantle can give the arrangement a livelier appearance.

Keeping the bouquet round or oval will probably present difficulties at first. So insert the flowers systematically, one kind after another, rotating the arrangement a quarter-turn each time you add to it.

The length to which you cut the stems is very important. You must take particular care with the long-stemmed flowers which will determine the final contours.

In order to obtain a livelier effect, rather than a compact arrangement, the bouquet can be built up in layers. Keep to the order of work as described above, but use more filling material. The flowers mentioned in stage 3 should not be inserted among the flowers in stage 2, but should stand a little above them. To achieve this, the stems should be left longer.

Triangular arrangement

A bouquet placed against a background to be seen from one side only is called a triangular arrangement or a half-bouquet. The back is left flat while attention is concentrated on the front and sides.

Any flowering-arranging book will tell you that the height of a triangular bouquet should be one and a half times that of the vase. The basic shape is a triangle.

This type of arrangement is made in the same way with dried flowers as when using fresh flowers. It is, however, more difficult with dried flowers to achieve the lively, luxuriant effect possible with fresh flowers. The dry stems lack the suppleness and fresh green leaves which give a fresh bouquet its stylish appearance.

If you want to use dried flowers in a tall vase, it may be difficult to achieve the desired height. The stems often have to be lengthened with wire. The

Step-by-step construction of a triangular bouquet:

1 The final height should be roughly one and a half times that of the vase, working within an imaginary triangle
2 Creating the contours

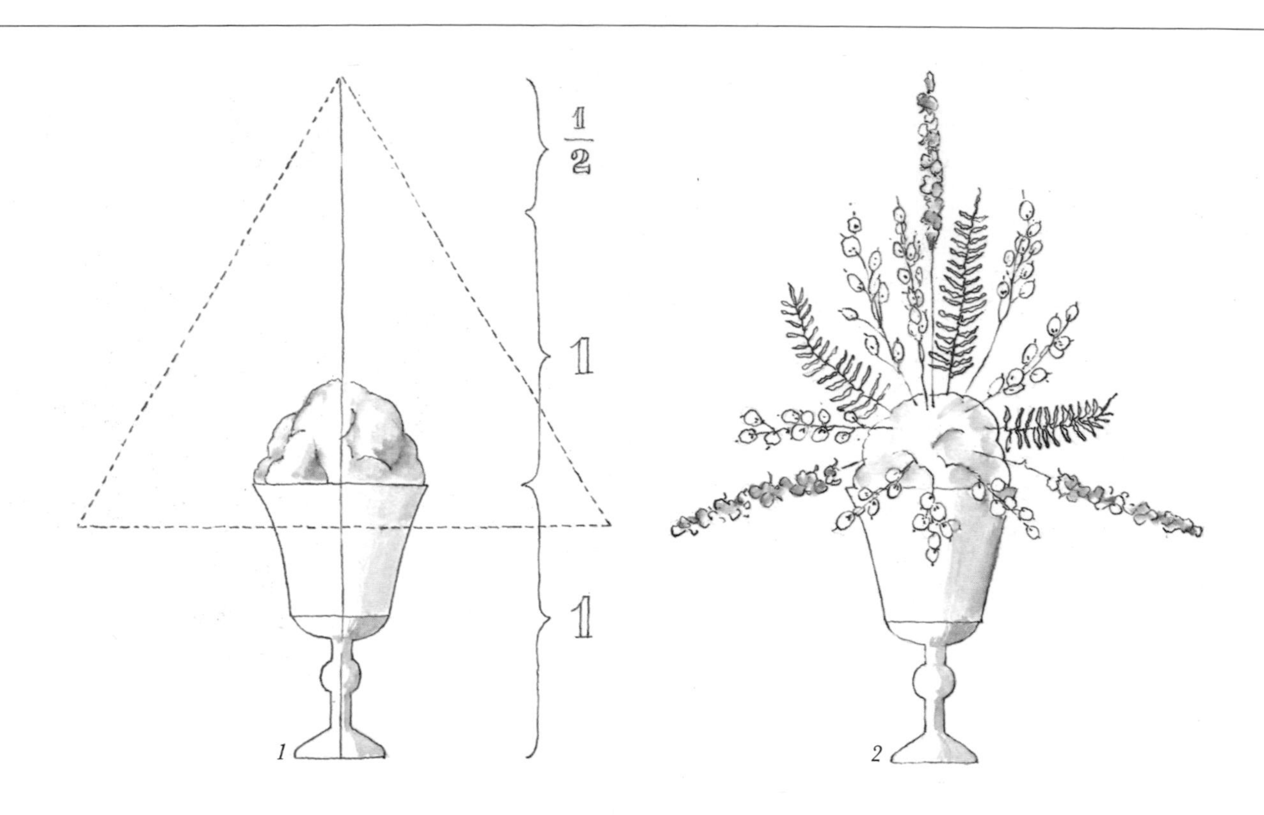

disadvantage of this is that after a while a forest of green wire becomes visible, which hardly enhances the natural look of the arrangement.

To avoid this, a large piece of Oasis should be left protruding above the rim of the vase (see diagram); the piece of Oasis should be well covered with moss. This makes it unnecessary for the stems to be so long, and also allows the heart of the arrangement to be filled more easily and the sides better shaped. It is important to remember that the side view of a triangular bouquet must also look attractive; it should not be a flat screen or a kind of fan, so the front and sides must be well filled too.

The order of work is as follows:

1 When the oasis has been secured and covered with moss, the shape should be determined by means of a few twigs, ferns, strong spike-shaped flowers or seed-heads. It should represent an imaginary triangle running from the apex down to both lower corners.

The bouquet must be given depth, so don't simply insert these first twigs in the centre of the Oasis, but as far back as possible, so that the sides can also be well filled.

The depth of the bouquet is also created by inserting quite long material in the front of the Oasis. In the bouquet illustrated on page 64 this was done with larkspur, honesty and Lamb's Tongue.

2 You can now set to work with the filling material as described before. The area close to the Oasis should be particularly well filled so as to distract the eye as much as possible from the places where the stems have been inserted. You can use achillea, hortensia and blue and white statice to give some impression of height, to about half the height of the contours previously established.

3 The centre or heart of the bouquet is now formed with a couple of beautiful, eye-catching flowers or bunches of smaller flowers tied together. In a blue

3 Inserting the filling material
4 The completed arrangement

bouquet, these could be carline thistles, scabious and edelweiss.

While a Biedermeier bouquet needs to be worked systematically, with a half-bouquet you have more freedom in the placing of different flowers. But that does not mean, for example, that the left upper half can consist of delicate flowers and the right of all kinds of heavy material. The aim is to create balanced proportions without striving for perfect symmetry.

4 If you have completed these three stages and you are satisfied with the shape, you can continue inserting flowers to your heart's content! A place can now be given to all kinds of flowers which accentuate the colour scheme, or grasses and seed-heads which enliven the silhouette. As this style of arrangement will be placed against a background, the over-all silhouette is important.

When adding more flowers, always keep a centre point in mind, as the stems should not cross over each other too often.

5 The bouquet can best be finished off in the position in which you intend it to stand. The way the light falls can also be taken into account and may reveal errors which can then be rectified. Check that there are no unsightly gaps visible at the sides.

A few tips for making bouquets etc.

1 Never use too many dominant flowers. It is better to use so-called filling material; this term covers flowers or seeds that are not over-conspicuous in shape or colour, such as all species of achillea, tansy, varieties of anaphalis, hortensia, meadowsweet and the less striking varieties of statice. Using your material in the correct proportions will ensure that the important flowers, whether dried hanging or in sand, are displayed to their best advantage.

2 Consider variations in shape when selecting your plant material. Don't

use only round or feather-shaped flowers.
3 Try to make the arrangements a little stylish by building them up in layers. In general, place the smaller, lighter flowers higher in the bouquet and the fuller flowers lower down.
4 If you are making a really large arrangement, use the smaller flowers in bunches; rhodanthe, helipterum, lonas and helichrysum can all be used in this way.

Flowers in full bloom can be attractively combined with their buds.

5 Use delicate flowers or seed-heads last to reduce the risk of damage.
6 Never work with material which is too dry; it is better to hang it the day before in an unheated room or dampen it a little with a plant spray.

Helichrysum is very sensitive to damp; the stems soon become limp. To avoid this, mount the flowers on wire.

7 I would recommend that you remove a dry arrangement from time to time, to reduce the chance of becoming bored with it and to increase its life-expectancy. It should preferably be stored in the dark.
8 Although the final shape of a bouquet depends mainly on the container, the place where the arrangement is to stand must also be a determining factor. If it is intended for a table, you must be able to look at it from all angles. A round, Biedermeier-inspired arrangement would be the obvious choice.

For an arrangement to be placed against a wall or in front of an empty fireplace, the triangular style would be the most suitable.

Decorations for parties and special occasions

For weddings, birthdays, anniversaries and other special occasions such as Christmas and Thanksgiving which are celebrated with a party, the flowers and decorations will often be as important as specially prepared food and drink.

Most hostesses nowadays have little time for preparation on such occasions, and cut flowers quickly lose their freshness. Dried flowers are an excellent alternative, as they can be arranged well in advance and will continue to give pleasure after the event.

Weddings

Fewer and fewer weddings are now celebrated in the old-fashioned, romantic way, but for those who cling to this tradition, it is useful to have some ideas for decorations.

Dried flowers are ideal for decorating the house of the bride, with the added advantage that they need not be arranged on the day itself.

A large wall decoration on wire mesh, as described on page 58, is most attractive hung over a door and can also be made in an arch shape.

Bridal trees are also very festive. They are made as follows:

As a base, you will need a square piece of wood, cut to the size of the Oasis you intend to use and nailed securely to the end of a broom handle. Bricks of Oasis can then be attached to the wood with wire, binding it round securely in different directions so that the result resembles a parcel, roughly cube-shaped. The wood must be completely

Bridal tree with green and white dried flowers shown on page 69. The background consists of Hydrangea petiolaris, meadowsweet, anaphalis, Achillea millefolium and white statice. White pompom dahlias dried in sand, white acroclinium, large burdocks and carline thistles were then inserted. The arrangement was finished off with bleached ferns (purchased), Baby's Breath and honesty.

covered with Oasis, thickly enough to allow you to insert flowers at any angle – don't forget to add some to the underside of the cube too. The final rounded shape of the tree is most easily achieved on a cube base.

Fix the broom handle securely in a suitable pot or basket with a large lump of clay. The clay will also make the container heavier and the tree will be more stable.

The container should then be filled with moss and the broom handle wound round with ribbon to match the colours of the flowers which are to be used.

Cover the Oasis with moss, and then start to insert the flowers, as follows: first the flowers that make good fillers, then the spikes for the contours etc., as for a wreath or a hanging decoration. Don't be mean with this first layer. By working close to the base, less material will be needed; otherwise there will be large gaps and you will need too many of the best flowers, which should be added last.

Two trees look even better than one. A nice gesture after the celebrations would be to offer one of these trees to each of the mothers of the newly-weds as a souvenir.

You can, of course, make smaller trees. The Oasis balls obtainable from garden centres and florist's shops can be used for these, fixed to a bamboo stick. They also should be placed in a pot or basket filled with clay.

Biedermeier bouquets of dried flowers can be very attractive for bridesmaids, with the additional advantage that they can be kept afterwards; and matching wreaths for the hair are easy to make.

Small bouquets are made in the hand. Begin with the centre and make each successive circle with a different kind of flower. As each round is finished, it should be secured with florist's wire before starting the next round. For the centre, use flowers with their own stems; the later rounds are better made with flowers or bunches mounted on wire. They can be bent outwards a little to avoid making the bouquet too stiff. When it reaches the required size, the stems should be bound with florist's tape, an elastic green ribbon which florists use to make corsages. A Biedermeier frill, some ribbon and a bow complete the work. Garlands of flowers for the hair should be made in the same way as the garland on page 72.

A flower basket can also be filled with dried flowers, either with heads left over from the bouquets described above, or with heads which have been especially removed from the stems.

You can probably think of many more ideas. Large hanging decorations, perhaps using a combination of fresh leaves and dried flowers, can look very festive.

Corsages

It can be a nice gesture to give a beautiful corsage to all the ladies present on a special occasion. If you make a number of corsages in different colours, each lady will probably be able to find one to match her clothes.

Corsages can also be made well in advance.

Arrange a small, oblong bouquet on a fern, which acts as the base. Use first the spikes and then the rounder-shaped flowers.

The stems should be bound together very close to the flowers, with corsage wire. If you want to bend the smaller flowers out into an attractive fan shape, it is advisable to mount each one separately on a wire stem. This, of course, takes time, especially if you need to make a whole basketful. The stems should then be concealed with florist's tape which can be obtained from flower shops. A small bow to match the colour of the flowers neatly finishes it off. The corsage can be attached to the clothing with a small safety pin.

Cradle bouquet

A very small Biedermeier bouquet, perhaps made from flowers the colour of the cradle, is a delightful gift to celebrate the birth of a child. You could also work in some tiny copper bells, obtainable from craft shops.

Table decorations

Whether for a formal dinner party or an informal supper, an attractively decorated table is always a festive sight and shows the thoughtfulness and care of the hostess. In spite of advancing

This corsage contains pink acroclinium, helichrysum, sea lavender and (drawn separately) purple and white xeranthemum.

emancipation it will, in most cases, be the hostess who has to take care of everything, including shopping, preparing the meal, and decorating the table; and when everything is ready she still has to give the impression of having done it all in the twinkling of an eye.

Decorations made with dried flowers to match the dinner service and/or tablecloth cannot only be made well in advance, they can also be used more than once.

You can, for example, combine a garland of flowers, which can be placed in various positions on the table, with one or more candlesticks decorated with the same flowers.

There are other ways of using garlands too. You can use one to brighten up a straw hat or to give a dull lampshade a totally different appearance.

A chair for a birthday child can be garlanded and even a bicycle can be decorated, for a special occasion or a pageant, perhaps.

If the garlands are stored in boxes (safe from mice) in a dark, dry place, they will give you pleasure for years.

How to make a garland

First prepare the flowers, etc., which you are planning to use; make sure you have a sufficient amount of each variety. Cut the stems to size, untangle any long pieces of moss and mount anything which needs a stem on florist's wire.

Now take a thick piece of wire (0.8mm), which will act as a frame for the garland. Twist one end into a loop, into which you insert the first bunch, using for this a small corsage arrangement with a rather pointed flower as a base; this will give the garland a neat, well-finished appearance. Attach this bunch to the wire frame by means of very thin wire, preferably corsage wire.

Then continue to add flowers, etc., one after the other, attaching each item securely with fine wire wrapped very firmly round both the stems and the thick wire frame. When the fine

wire comes to an end, start with a new length. It is not essential to work it in neatly as in embroidery work.

When you reach the end of the wire frame, bend the end into a loop. A new piece can then be threaded through the loop, bent over and securely twisted. In this way the garland can be made as long as you want, attaching more flowers to the new piece of frame wire.

If the flowers are not too dry, their own stems can generally be used.

When the garland reaches the required length, another bunch should be attached, as at the beginning, to finish it off. Place this bunch lying in the opposite direction to the main part of the garland, taking care to tuck the stems under the last flowers on the wire, and attach it securely. If you want to turn the garland into a wreath, this last bunch is not necessary; the first bunch of the garland will cover the final stems.

A Christmas ball (described on page 75).
Start with a ball of Oasis covered with Icelandic moss.
Larch cones, artificial Christmas apples and white helichrysum
can then be inserted. The ball is finished off
with ribbon bows, Baby's Breath and dried cineraria leaves.
When hanging it can be decorated with holly or fir.
The rest is left to the spider.

Christmas decorations

However Christmas is celebrated, few people can resist the special spirit which belongs to this time of year. There is scarcely a home where people do not make some effort to cheer themselves with Christmas decorations.

The approach is naturally enough always very personal. One person will make do with a simple tree or a branch here and there, another may festoon the whole house; but however it is done, ideas for Christmas decorations are usually very welcome.

If you store the decorations described here in a dark, dry place, they will give you years of pleasure. No-one will get tired of them as they will be seen for only a week or two each year. Furthermore, some of these decorations are so strong and easy to transport that you can take them with you if you are celebrating Christmas away from home.

Decorated candlestick, using mastic (see below). Sand-dried galtonia (the white bell-shaped Summer Hyacinth) is combined with hortensia, helichrysum, honesty, white rhodanthe, Baby's Breath and the attractive feathered leaf of cineraria.

Candlesticks

When decorating a candlestick, the best aid is mastic, a type of pasty green clay which does not harden completely. It has a rather tough consistency from which fragile dried flower stems cannot always be easily extracted – a relatively unimportant disadvantage when compared with the fact that it can be moulded into any shape. For a simple, straight candlestick, a sausage shape can be rolled round the candle. When dealing with candelabra, two small balls of the mastic can be placed where the branches separate. These are just two suggestions.

When decorating candles, bear in mind that dried flowers are very inflammable. It is best to use tall candles.

When the mastic has been firmly stuck to the candle, decorating can begin. Cover the mastic with a little moss and then insert flowers. Since most small flowers or bunches will be mounted on wire, they can be bent in any direction required.

If space is limited (which will certainly be the case with any normal-sized candlestick), each flower must be inserted in exactly the right position, facing in the right direction.

Larch cone mounted on wire

Advent or Christmas wreaths

Wreaths can be used in various ways for seasonal decoration. They can be laid flat, or hung on a wall or from the ceiling. An advent wreath is usually hung horizontally on four ribbons tied together in the middle and hung from the ceiling with a bow.

Fresh fir greenery is very attractive for wreaths, but moss and dried flowers are also good. The advantage of the latter is that they do not drop and can be used again the following year.

The basis for the wreath is made from Oasis in the same way as described for the flower wreath on page 60.

Moss plays a really important part here; it is not just a base but the background of the arrangement, which will remain visible when the wreath is finished.

The decoration of the wreath depends on what colour of moss you use; for example, white helichrysum, larch cones, plenty of Baby's Breath and bows of off-white ribbon make a lovely combination with green moss. Or try a composition of very pale gold spotted nicandra and bunches of white xeranthemum with a few gold-sprayed water plantain seed clusters and gold ribbon.

Grey moss can also be very cheerfully decorated with some of those artificial red Christmas apples combined with alder berries and bunches of white xeranthemum. The arrangement can be finished off with red ribbon.

There are two ways of putting candles into a horizontal wreath. You can buy special candle holders which can be inserted in the Oasis, or you can make the holes in the Oasis yourself, using an apple corer, and place the candles in them.

Again, remember that dry material is highly inflammable.

Small bows can be made beforehand and pressed firmly into the Oasis with a piece of bent wire, with the ends of the bow well separated.

Christmas apples are available in all colours and sizes. They are usually mounted on a steel wire and are very easy to use.

If you want to use beads, thread corsage wire through each one and give it a few twists; the wire can then be inserted in the Oasis.

A wreath for the door can be completely covered with nuts and pine-cones, possibly combined with large seed-heads.

There are, of course, many more ways of giving a wreath a festive look.

Carline thistles or even downy, gone-to-seed thistles (see page 46) give an attractive, shiny effect.

The best ideas will usually occur to you once you have started work.

Christmas trees and Christmas balls

Like the wreath, these can be made on a base of Oasis. Balls and cones of Oasis can be purchased from a florist. These should be covered with moss until the Oasis can no longer be seen, followed by more or less Christmassy flowers and seed-heads, such as helichrysum and xeranthemum. If the ball is to be hung, pierce it with wire, make a hook at the bottom and pull it into the ball. Bend another hook or loop at the top to which a ribbon can be attached (see illustration on page 76).

When decorating these shapes, take care that the work is even and that the material used, such as larch cones, is of approximately the same size.

Other suitable materials are artificial apples and berries, small toadstools, beads, and bows. Very small balls to add to a Christmas tree can be made by completely covering the smallest size ball of Oasis with white xeranthemum flowers, with a few artificial holly berries in between and a red ribbon by which the ball can be hung up.

(left) A hanger for a hanging arrangement made by pushing a strong wire through the Oasis base
(right) A Christmas 'tree' made from dried Water Plantain

When decorating a cone, make sure that the shape is retained. The flowers, pine-cones, etc., you use should decrease in size as you work towards the top. These cone arrangements can simply stand by themselves. Very small cones or balls can be made into miniature trees by piercing them with a stick and standing them in a pot in a clay base which can then be covered with moss.

Large cone-trees are easy to transport, and if they should be tipped over, the damage is generally only slight.

A Christmas surprise

A Water Plantain Tree
(illustrated opposite)
Carefully hung and dried water plantain can be made into a miniature Christmas 'tree'. Take care not to let the branches become entangled – they are brittle and will break. If the plant is large, it is advisable to cut off the top, keeping two layers of side branches. It can be placed in a base of clay wrapped in aluminium foil which is then covered in moss.

The 'tree' should now be lightly sprayed with gold paint to give the delicate branches and seeds a slight shine. It can then be decorated with artificial berries, transparent beads and bows.

If the tree is not too large, it will make an attractive table decoration since its delicate shape will not get in the way of conversation.

For this dry bouquet from the herb garden,
tansy, hop and Lady's Mantle were used as a foundation.
The contours were created with wormwood (1), hyssop (2), lavender (3), and
sorrel (4), followed by wild chamomile (double variety) (5)
and wild marjoram (6). Finally the attractive seed-heads
of chervil (7), dried yellow rosebuds (8) and the
beautiful feathery seed-pods of borage
or cucumber were inserted.

Herbs

For centuries herbs have played an important role in our lives. Thousands of years before Christ, they were already being grown for their medicinal, culinary and cosmetic properties.

Witchcraft was associated with herbs, and magical love potions were brewed from them. Herbs were also important in mythology and astrology.

In the first century AD, the Greek physician Dioscorides wrote *De materia medica*, a treatise on the use of medicinal herbs which was to be used for many centuries. And the discovery of old papyrus scrolls revealed that the ancient Egyptians recognized the medicinal value of thyme, caraway and garlic.

The Romans also used herbs extensively: they used to scatter lavender, roses and saffron on the floor to give their houses a pleasant fragrance. They used the same herbs to perfume their baths. Their knowledge of herbs spread throughout Europe as a result of their military and colonizing campaigns.

The cultivation of herbs reached its height in the walled gardens of mediaeval cloisters and castles and, in addition to herbs, vegetables and fragrant flowers such as lilies, roses and violets were also grown.

In the past century, developments in the field of chemistry and improved hygiene have caused a considerable decline in the use of herbs but there has fortunately been a revival of interest in recent years. People once again appreciate the pure fragrance of herbs instead of artificial flavours and aromas.

For herb enthusiasts, this revival of interest means that the seeds and plants of all kinds of herbs are relatively easy to find. All kinds of flowers and leaves grown in the herb garden can be successfully combined in a bouquet of dried flowers.

Since the herb family is so extensive, we shall limit outselves here mainly to culinary herbs and to methods of drying them for the store cupboard or to add to decorations.

Picking and drying

The time to pick depends on the purpose for which the herbs are to be used so instructions for picking and drying are divided into three categories:

Flowers

The flowering herbs which provide colour in your herb garden usually dry very well. The following are particularly suitable: marjoram, chives, bergamot, varieties of mint, valerian, hyssop, lavender, double chamomile, wormwood and tarragon.

Pick the flowers just before they reach full bloom. They will open further during the drying process. Chives, especially, should not be picked too late. Sprigs of wormwood and tarragon, however, must be picked in full bloom.

Dry the flowers tied in bunches and hung up in a dry, well ventilated place (see Drying by hanging, page 13).

Seed-heads and seed pods

If you want to use the seed-heads in decorations, you should pick them before the seeds have fully ripened. For culinary use, the seeds must naturally be picked when completely ripe. Angelica (comparable in appearance to giant cow parsley), dill, lovage, coriander, rue and chervil have the most decorative seed heads.

Borage also has beautiful grey, hairy seed capsules. For drying, they must be picked while the last blue flowers are still in bloom.

Sorrel soon disintegrates. Pick when the seeds turn red.

Sage which has finished blooming has an attractive shape and colour and is ideal for use in arrangements.

Seed pods, and heads should also be tied in bunches and hung up to dry.

Leaves

The leaves, singly or on sprigs, of rue, wormwood, sage and bay are best lightly pressed for later use (see page 15).

For culinary use, the important thing is to preserve the aroma. Many herbs contain the maximum amount of aromatic, volatile oils when the first flower-buds are on the plant.

Pick the herbs when the dew has just dried and before the sun is at its highest: too much warmth from the sun causes evaporation of the volatile oils.

Harvest only the tops of the plants so that some leaves are still left on the stems, allowing the plant to shoot again.

Chervil, parsley, celery and lovage should be picked quite early, while the leaves are still tender.

The shrub-like herbs such as sage, thyme, marjoram, rosemary and bay retain their fragrance extremely well.

Tie these herbs in fairly small bunches and hang in a dark, dry, well-ventilated place. When they are thoroughly dry, strip the leaves off the stems and crumble them. Store the crumbled herbs in well-sealed pots in the dark.

Leaves that take longer to dry, such as dill, fennel, parsley, celery, chives, lovage, sweet basil and chervil retain their fragrance much better if dried quickly in the oven. Remove the leaves from the stems and place them well spread out on a baking tray. Chives should first be chopped.

The oven temperature must be no higher than 125°F (50°C) otherwise the volatile oils will evaporate. During the drying process, the oven door should be left ajar, to give the moisture a chance to escape. Remove the leaves as soon as they feel dry and can be easily crumbled.

Store the herbs as described above.

Decorations with dried herbs

Although herbs can be very attractively combined with ordinary dried flowers, you could also try a bouquet made entirely of herbs. It will be rather less colourful but full of the smell of summer.

Here are a few ideas for decorations which include material that can also be used either for flavouring food or in the bathroom.

A culinary wreath is composed of herbs, herbal sachets and herb flowers used in cooking.

A bathroom wreath can be made in the same way, but using different fragrances.

This procedure should also be followed for an Italian garland made up of a string of garlic decorated with marjoram and chive flowers, shallots, sprigs of thyme and basil, and possibly a ribbon.

Another variation is a plait (braid) of straw decorated with herb flowers, bay leaves and herbal sachets. You can make one using only fish herbs, or consisting of oriental spices combined with small red peppers and finally, for decoration, the dried flowers and seeds of coriander.

The technique for making a straw plait is described on page 89–91.

Culinary wreath

The basis of the wreath consists of a ring of Oasis covered with moss (as described for flower wreaths, page 60). For a culinary wreath, insert herbal sachets (see below), shallots and garlic all over the wreath, and decorate with marjoram flowers, dill or fennel heads, sprigs of thyme, tarragon, southernwood and pressed sage or bay leaves (these should be mounted on wire).

The filling for a herbal sachet is a *bouquet garni*: a mixture of dried herbs which go well together and which is left to infuse in sauces, stews, marinades and bouillon. The basic mixture is:

1 tablespoon dried parsley
1 teaspoon dried thyme
1 bay leaf
2–3 peppercorns

You can, of course, add to this mixture according to the dish and your taste.

Crumble the dried herbs and put a tablespoonful of the mixture in a piece of cheesecloth or muslin 15cm square (6 × 6ins). Gather the edges together and secure tightly with wire, leaving a piece of wire to insert in the wreath. A clove can now be pushed into the centre of the herbal sachet.

Other good mixtures of herbs:

with beef: lovage, celery, marjoram
with pork: sage, marjoram, rosemary
with poultry: tarragon
with lamb: rosemary
with Italian dishes: basil, marjoram (even better is oregano or wild marjoram)
with fish: chervil and fennel or dill and a very little southernwood or a piece of lemon peel
in marinades: more thyme, more peppercorns, juniper berries and, with fatty meat, rosemary

Since these extra herbs generally have a strong taste, you should use them sparingly except for a *bouquet garni* for marinades which can be a little stronger.

The herbal sachets are the most important ingredient of the culinary wreath as they are there to be used. Make rather more than you need. If you use them regularly, the wreath will soon begin to show gaps which you can then fill up again with the spare sachets. Store them in glass jars.

Bathroom wreath

The same method is used for a bathroom wreath, but the combination of herbs and flowers will be completely different.

Sachets for a bathroom wreath should be filled with fragrant flowers and herbs. They can include mixtures for the bath water or for rinsing hair. Lavender or marjoram flowers, rosebuds, Baby's Breath and the pressed leaves of the fragrant varieties of geranium (pelargonium) can be used for decoration. Although they all belong to the herb family, these can also be found in the perennial plant border.

You can experiment with herbs and flowers and invent your own blend.

Two good blends are:

Lavender-scented mixture

plenty of lavender
bergamot
a little mint
fragrant rose petals
rosemary
dried orange rind

Lemon-scented mixture

lemon balm
lemon geranium
bergamot
a little southernwood
dried lemon rind

A bath sachet should be hung under the hot water tap while you are running the bath. A stronger fragrance can be obtained by first infusing the sachets for an hour in boiling water. Leave the lid on the pan and remove from the heat.

Rosemary and chamomile are very good for the hair. Blondes should use chamomile sachets and brunettes rosemary, to give a fairer or darker shine respectively.

Leave the sachets to infuse in boiling water and use this infusion as a final rinse. The use of rainwater will give your hair a shine as well as a pleasant smell.

Single sachets of different fragrances are excellent for hanging in the bathroom and in bedroom cupboards, for example.

Sachets filled with anti-moth herbs provide the solution to a cupboard infested by moths. The appropriate herbs are tansy, bergamot, santolina and southernwood. Hang an arrangement, filled with sachets of these herbs and decorated with their flowers, inside the cupboard door.

Once you start drying herbs you will undoubtedly think of even more ideas.

A hat or a lampshade can be decorated with a garland of dried flowers (see page 71–2).

(below) Different varieties of grass suitable for pressing
(opposite) Wild flowers for pressing:
1 Sorrel 3 Pink-tinted Achillea millefolium
2 Hogweed (Cow Parsnip) 4 Shepherd's Purse

Pressed flowers

Everyone will at some time have pressed flowers or leaves in a thick book. Anyone lucky enough to have found a four-leafed clover will undoubtedly have kept it in this way. The method of drying flowers by pressing them is quite old. There are still books dating from Victorian times in which the entire Alpine flora has been preserved; and old herbaria from all parts of the world are still in existence.

Making a flower press

To avoid disappointment with pressed flowers, I recommend the use of a proper flower press. The drying proccss is much quicker and the resulting pressed flowers are of a better quality. A small press can be bought in handicraft shops; but if you want to dry a lot of flowers in this way and need a larger press, it is quite easy to make one yourself.

You need:

Large sheets of blotting paper folded double – A3 (297 × 420mm), or approx. 11 × 16ins, when folded is a good size.
2 pieces of strong plywood approx. 50mm (2ins) larger all round than the blotting paper
old newspapers
4 bolts ± 7mm diameter
4 matching wing nuts
4 matching washers

Holes should be drilled in the corners of the plywood to take the bolts.

Alternate layers of newspaper and folded blotting paper are placed between the two pieces of plywood, and the press is closed by pushing the bolts through the holes from below, placing a washer over each one and screwing down the wing nuts.

Pressing

Grasses, mosses, flowers and leaves should be gathered on a fine, sunny day. Make sure that everything is thoroughly dry. The material collected should be placed in the press as soon as possible since if it has withered it will emerge crumpled after pressing.

Start with simple flowers such as violets, single hortensia florets and buttercups. If you prefer wild flowers and plants, pick sparingly and only from thickly covered verges.

When working with pressed material, you will soon discover how much can be done with a few flowers, ferns and grasses. In principle any flower can be pressed, but the colours do not, unfortunately, always remain true.

When you are ready to start pressing, take a double sheet of blotting paper and lay it open.

Flowers generally have a rather thick centre which you should gently squeeze flat between your thumb and index finger. Then place the flower face downwards on one half of the blotting paper, making sure that the petals are not folded or crumpled.

Grasses and mosses may need to be thinned out a little so that they are not too 'busy' and the attractive shape is not obscured.

Leave sufficient space between the

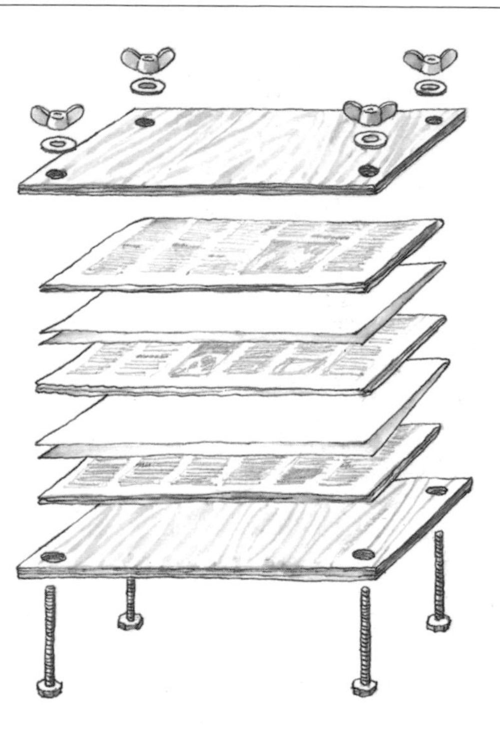

flowers and place material of approximately the same thickness on one sheet, or the pressure will be uneven. If the hearts or stems are too thick, they are better cut off and dried on a separate sheet.

Check that everything is in the correct position and carefully fold the other half of the blotting paper over.

Place a folded newspaper the same size as the blotting paper at the bottom of the press and lay the folded blotting paper containing the flowers on it; follow this with another folded newspaper, more flowers, etc. Close the press and screw down the nuts.

If you put very thick material between the layers of blotting paper, use a thicker layer of newspaper on either side to avoid pressing the bumps through to other layers.

In a centrally heated house, it will take about a fortnight for the flowers to dry thoroughly.

You can open the press to add new layers of material during this time, but be careful not to disturb the first layers.

The first sheets of blotting paper can be removed after two weeks. If the centres of some flowers adhere to the paper, gently loosen them with a thin knife.

Quick method of pressing

For a quick method, the flowers should be laid in the same way between blotting paper. Place a layer of newspaper on a hard surface, followed by the blotting paper and then another layer of newspaper. Now iron all over – the iron should be just hot enough to heat through the whole pile.

Storing pressed flowers

Since flowers and leaves soon curl up in a damp atmosphere, it is best to store them in an old telephone directory or between old magazines. Put them away carefully in a safe place to avoid the whole pile being thrown away, by mistake, along with old paper.

Gluing

Whatever plans you have for using pressed flowers, they should always be glued to a background. This should be done systematically. Once you have selected the flowers to be used, arrange them with tweezers or a thin knife on the chosen background. When they have been arranged to your taste, carefully smear the back of each flower and stem with a thin layer of glue. Don't forget that they are fragile. A minimal amount of glue is necessary. Ordinary transparent photograph glue, sold in a tube or bottle with a nozzle, can be used accurately without leaving too many marks.

When working on a varnished or painted background, you can use a little varnish smeared on the back of the flower with a fine paintbrush. This is ideal for very fine grasses and feathery leaves.

When all the flowers have been smeared with glue or varnish, place them in position in the arrangement chosen.

Work carefully, since the glue will leave marks if you try to make corrections. When everything is in place, lay a sheet of blotting paper on top, followed by a heavy plank or a thick book to counteract any curling.

Leave for a day. The glue will then be thoroughly dry, and you can finish off the arrangement in one of the following ways.

Flower pictures should be covered with a sheet of glass and a frame. Articles to be used are best coated with natural matt varnish. Greetings or menu cards can be covered with adhesive plastic film.

Suggestions for using pressed flowers

Even if you are not particularly impressed with your first efforts they will in any case have been good practice. Making greetings and Christmas cards or gift labels can provide considerable experience.

Decorated matchboxes, candles, trays, birthday calendars, place-mats and bookmarks make delightful gifts. Protect articles of this kind with a layer of varnish or adhesive plastic film.

An Easter egg tree is inspired by a Swedish custom. A few weeks before Easter, pick an attractively-shaped branch of birch and leave it to sprout in a vase. The fresh green branch can then be decorated with Easter eggs. First blow the eggs, and then stick flowers on them and varnish them. These eggs can be kept for years if stored carefully in the dark.

Once you have mastered the gluing technique, you can attempt a picture. Choose the colour and the material for the background very carefully. It could be painted or coloured card or it could be made of coarsely-woven linen or real silk mounted on card.

The colours of the picture could be arranged to harmonize with the colours of the room where it is to hang.

Try combining pressed flowers and leaves with ordinary dried flowers.

This straw cross was made in the same way
as the Tree of Life described on page 91, using wheat (1),
barley (2), rye (3), and oats (4).

Straw

Straw is splendid material to work with. Many kinds of decorations and useful articles can be made from it.

Straw was also used in this way in days gone by. In many European countries there were old traditions such as placing straw decorations on the last sheaf of corn at the end of the harvest. In Sweden many things are still made from straw, especially at Christmas. In England making 'corn dollies' is a traditional country craft.

Straw is unfortunately not so easy to come by these days. Agricultural machines cut the stalks to pieces and bale them immediately. You can try asking a friendly farmer to reap a little for you before the harvest. He may even allow you to cut a little for yourself.

All kinds of straw can be used. Rye straw has the longest and most pliable stalks and is therefore particularly suitable. Other varieties, such as oats, barley and wheat, can also be used. Wheat ears are very decorative. Different kinds of grasses, picked by the wayside, can be combined with the straw.

Before the straw can be used, each individual stalk must be peeled; that is, the dull protective skin around the stalk must be removed. This is a lengthy, boring business and is best done immediately after picking. The stalks are then very pliable and do not break so easily, and the mess is kept outside.

Only the upper part of the stalk is necessary for more delicate work; it can be snipped off above the last node. The protective skin will then slide off easily.

Various techniques, originating from different regions, are employed in making straw decorations. Coarse plaits (braids) made from bundles of straw are sometimes used, but fine spiral basketwork can be made from single stalks. Christmas figures and animals can also be made from straw.

Only relatively easy suggestions will be given in this book. Specialized books on straw-craft are available for anyone who wants to find out more or to become an expert in the use of this material.

A corn stalk is divided into three parts with a node between each section: the lower part, the middle, and the upper part which includes the ear.

The upper part is the finest and therefore ideal for small decorations and plaits (braids), and 'staircases'.

The middle section is somewhat coarser in structure but is very suitable for binding Christmas stars and small straw figures, for example. The straw sold in handicraft shops resembles this part of the stalk.

The lower part has no special use but is included when the whole stalk is to be plaited (braided). The thick node between the middle and lower sections may be a problem, as it tends to protrude, giving the plait an uneven appearance.

As the whole stalk can be used for large plaits, the pieces which remain when the upper part has been cut off to use for smaller decorations can be worked into your plait bundle.

All kinds of things can be made from these large plaits: wreaths, hearts, frames for mirrors, and a base for a herbal decoration such as that described on page 81.

Oat straw is particularly attractive if some ears are left on the stalks. When the plait has been completed, a few fine oat grains can be seen here and there, giving the plait a livelier accent. The stiff ears of wheat, rye and barley are less suitable for this purpose.

Before you start work, the straw must be soaked. The bath is the best place to soak long stalks, because they must lie flat. Never soak more than you intend to use at one time. Straw which is repeatedly soaked and then dried loses some of its colour and shine; and straw which is too dry breaks when you start to plait it.

It is essential to use strong thread to bind the straw. Button-hole silk or linen thread is the best, in the same colour as the straw. The thread should be bound tight enough to crack the straw.

A Christmas wreath made from straw and decorated with hearts and balls made from coloured card

For a large plait made from soaked straw, it is better to have half the stalks lying in the opposite direction to the other half (see diagram). As the lower part of the stalks is thicker than the top part, the bundle would taper to a point if the stalks were all lying the same way.

Bind the bundle at the top and divide it into three equal portions or 'ropes'.

When plaiting, always fold the outside ropes alternately over the centre one, so that the curves on each side lie in the same direction. The plait must be firm and even.

To add new straw, place a number of new stalks (depending on the original thickness) along the centre rope and weave them in. Take care that you add an equal amount to each rope as it comes to the centre. The evenness of the plait is easily spoilt by adding too much or too little new straw when making a long plait. If any ends protrude where new material has been woven in, they can be cut off later with sharp scissors.

If you want a plait with a closed top, you should proceed as follows. Bind the bundle tightly in the middle. Then bend it in the centre into two equal parts and re-divide these into three ropes and plait them as before, adding extra material as described.

When the required length is reached, the lower ends should be tightly bound with thread. If the plait is to be made into a wreath or heart, don't cut the ends off until the required shape has been formed.

A plait with a closed top is better for a wreath. Bend the spiky end round and place it inside the closed top and stitch fast with a needle and thread. An untidy join can be camouflaged with a ribbon or a bunch of ears or grasses.

The wreath illustrated was made from a joined plait. Four pieces of red ribbon were added to hang it up and it was decorated with balls and hearts made from red card. Cut out three identical circular or heart-shaped pieces of card for each ball or heart. Fold each in half and fasten them together down the middle with a long piece of red cotton, then open up the folds. Sew each decoration firmly to the wreath.

A straw wreath can also be decorated with bunches of different ears, grasses and dried flowers, sewn firmly to the straw, or with larch cones and alder berries attached with wire.

Straw decorations are not only suitable for Christmas or Easter. A straw heart, for example, is attractive the whole year round. To form a heart shape from a plait, tie the two ends of the plait together lying in the same direction. The join can be embellished with grasses, ears, cones or ribbon. The point of the heart can be accentuated by threading a piece of strong wire into that part of the plait and bending it to shape.

Small plaits and two-straw 'staircase' plaits

The upper, fine part of the stalk is used for small plaits and 'staircases'. Before plaiting, remember to soak the straw

Making as straw plait (braid):

1 The corn stalk is in three equal parts divided by two nodes.
2 For a large plait, the bunch of stalks should be divided in half and laid together in opposite directions to ensure equal thickness.
3 Tie the heads together at the top and divide the bundle into three equal 'ropes'. Always add new straw to the centre rope.
4 Starting a plait with a closed top.

(below) A Tree of Life

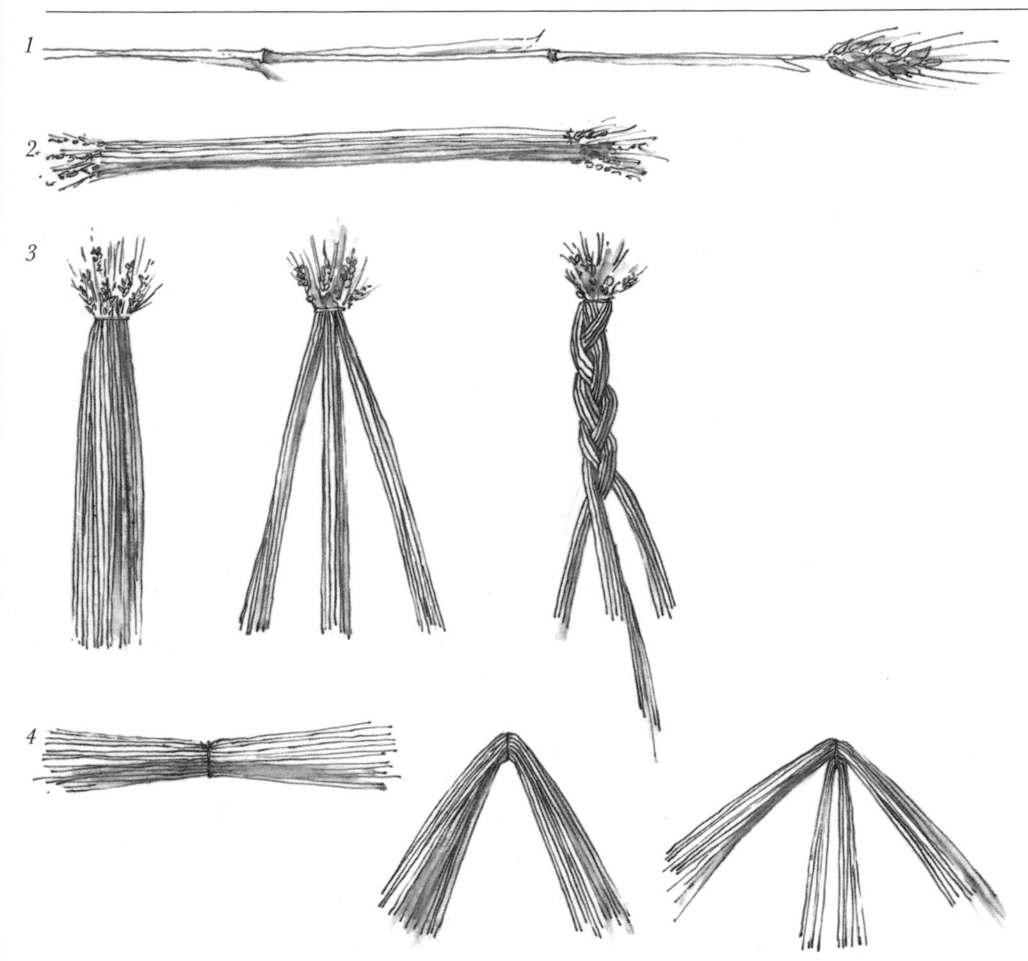

The Tree of Life

A 'tree of life' can make a gift suitable for any occasion and its symbolism will undoubtedly give great pleasure. The method for making it is as follows:

Bind a combination of oat and wheat stalks tightly together just below the ears. If the thread is tied tightly enough, the ears of corn will stand out a little. For a large tree, this vertical stem or 'trunk' should be strengthened with thick florist's wire.

The two horizontal branches are of smaller bunches of straw tied firmly with the heads lying in both directions, so that there are ears at both ends of the bunch. These branches are then placed in position between the stalks of the main trunk, at right angles to it, and securely tied in position with strong thread.

The hearts, each made of two plaits of exactly equal length, are also bound in position with thread and can be decorated with extra tufts of corn ears.

first to make it pliable. To obtain an even plait, choose three stalks of equal thickness.

As with the coarse plait, the stalks should be folded inwards alternately to achieve a good, even result. Weaving in extra straw is done in the same way. Any spiky pieces sticking out should be cut off when the plait is complete. Long plaits can be made in this way.

For the familiar two-straw 'staircase' plaiting, two stalks of equal thickness are folded over each other repeatedly at an angle of 90° until a kind of spiral 'staircase' is formed.

This kind of plaiting can be used in a number of ways: decorations for the Christmas tree, quickly made in the shape of hearts, balls and cracknels; napkin rings; and hairbands, enlivened perhaps with hare's tail grass, pearl grass, oats, larch cones, alder berries or artificial berries and toadstools – these are just a few suggestions.

Fixing bricks of dry Oasis together by means of strong wire

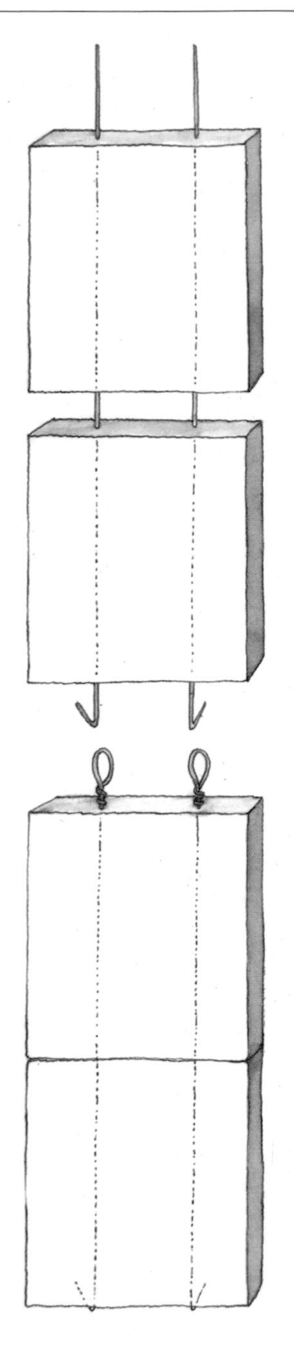

Materials and equipment

Oasis: soft but rigid green foam into which stems can be pushed (can also be used for fresh flowers as it absorbs water); available from florists and garden centres in bricks of various sizes.

Bouquet holder: soft Oasis in a plastic frame, used as a base for hanging arrangements; available in different sizes.

Dry Oasis (Oasis Sec): hard foam that does not absorb water, specially made for dried flowers; available in bricks, balls and cones of different sizes.

Wire: green florist's wire in different lengths and thicknesses; the most useful are the thin, very pliable (0.4mm) wire and the stronger 0.6mm and 0.8mm varieties.

Staples: strong wire staples are used to hold moss in place; 0.8mm wire bent into the shape of a staple can also be used.

Mastic: a type of pasty, green modelling clay, used when decorating candlesticks. Can also be used for wide vases and dishes in which it is difficult to secure Oasis: put some mastic in the bottom and then fix the Oasis to it by inserting long pieces of 0.8mm wire through the Oasis into the mastic. The clay does not dry out; it is rather stiff, and thick stems may break off when inserted. Alternatively, use Plasticine.

Flower prongs: plastic prongs or spikes with an adhesive base are available from florists for securing Oasis to a flat plate or dish.

Clay: use ordinary clay (obtainable from toy-shops or handicraft shops) to make a heavy base in a basket to be used for an arrangement.

Florist's tape: elasticate ribbon used on small Biedermeier bouquets and corsages, to cover the wire and stems.

Pincers: sharp pincer or pliers can be used for cutting wire.

Scissors: a pair of heavy, sharp kitchen scissors for cutting flower stems.

Knife: to cut Oasis to the required size.

Hair lacquer: flowers or seeds-heads which easily disintegrate can be sprayed before use with hair lacquer. Sand-dried flowers can also be fixed in this way.

Gold paint in an aerosol spray can be used for Christmas decorations.

Paper ribbon: obtainable from stationers or shops dealing in window-dressing materials, in a wide variety of colours, for embellishing corsages, wreaths and bridal trees.

Moss: reindeer moss and sphagnum moss are used for covering Oasis before other material is added. They can be ordered from florists' shops and garden centres.

Sand: for drying flowers, use silver sand, obtainable from garden centres and some hardware stores.

Silica gel: silica gel crystals can be bought from any good chemist or drugstore (though it may be necessary to order in advance).

1 Brick of dry Oasis
2 Cut out required shape with sharp knife, using plate as marker
3 The inner circle for a wreath can be marked with a smaller plate or saucer
4 Round off vertical edge with a knife
5 Cover Oasis with reindeer moss fixed with staples
6 Staple for fixing material to Oasis
7 Bouquet holder
8 Creating and oval contour
9 Side view
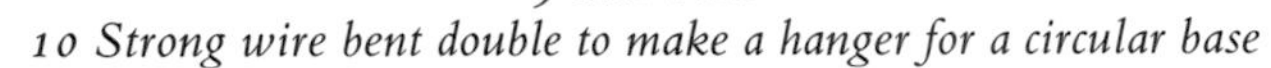
10 Strong wire bent double to make a hanger for a circular base

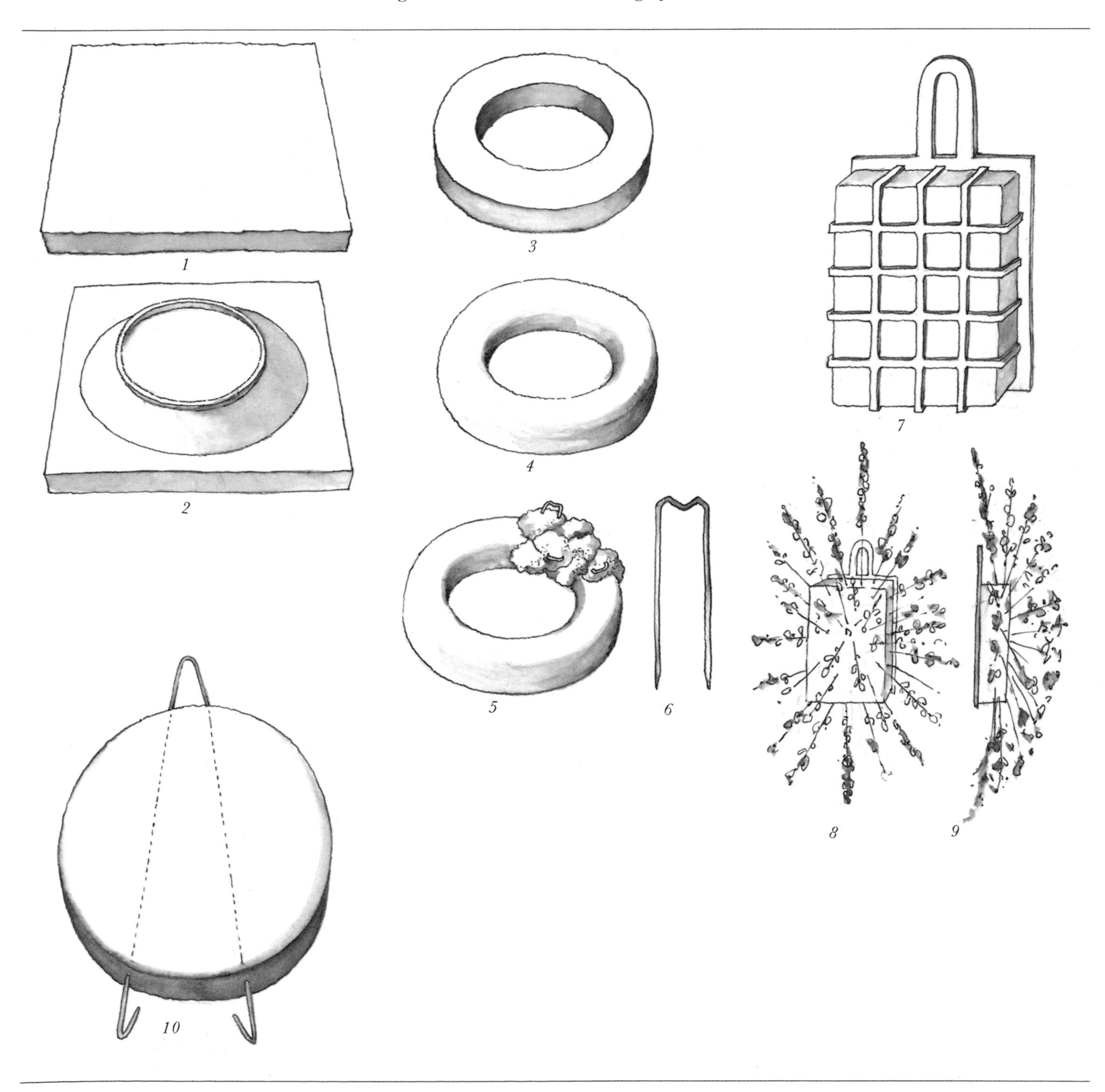

1 Mounting a leaf on a wire stem
2 Mounting a flower on a wire stem
3 Smaller flowers bound into bunches with florist's wire, which also acts as a stem

Latin plant names

English plant names